Ecology and Human Organization on the Great Plains

INTERDISCIPLINARY CONTRIBUTIONS TO ARCHAEOLOGY

ECOLOGY AND HUMAN ORGANIZATION ON THE GREAT PLAINS
Douglas B. Bamforth

HOLOCENE HUMAN ECOLOGY IN NORTHEASTERN NORTH AMERICA
Edited by George P. Nicholas

THE PLEISTOCENE OLD WORLD: Regional Perspectives
Edited by Olga Soffer

Ecology and Human Organization on the Great Plains

DOUGLAS B. BAMFORTH
University of Nebraska
Lincoln, Nebraska

PLENUM PRESS • NEW YORK AND LONDON

Library of Congress Cataloging in Publication Data

Bamforth, Douglas, B.
Ecology and human organization on the Great Plains.

(Interdisciplinary contributions to archaeology)
Bibliography: p.
Includes index.
1. Indians of North America—Great Plains—Antiquities. 2. Paleoecology—Great Plains. 3. Grassland ecology—Great Plains. 4. Great Plains—Antiquities. I. Title. II. Series.
E78.G73B355 1988 978′.01 88-22520
ISBN 0-306-42956-X

A Division of Plenum Publishing Corporation
233 Spring Street, New York, N.Y. 10013

Printed in the United States of America

Preface

Cultural–ecological research is one of the central themes of modern anthropology, and the past two decades or so have seen a tremendous increase in the methodological and theoretical sophistication achieved by such research and in the depth of the insights it has given us into human ways of life. However, much of this increased sophistication derives from an intense emphasis on a fairly narrow range of topics, particularly in archaeology: we now know a lot about resource selection, technology, and settlement/mobility patterns, but we are not as far along in understanding either how variability in these areas affects the other things people do or how cultural–ecological relationships may affect these other things more directly.

This study is an attempt to expand upon our present state of knowledge and address the ways in which certain aspects of the natural environment which a society exploits affect the way in which that society is organized. The justification for such an emphasis is simple: distributing a population across the landscape, choosing and procuring food, obtaining the materials needed to make tools, making the tools themselves, and getting them to the people who need them once they are made all require human beings to make decisions and act on them as a group, and the way in which those humans organize themselves provides the mechanism by which such decisions are made and carried out. The material conditions in response to which decisions are made and under which they must be carried out should thus be linked to this mechanism.

The region examined here is the North America Great Plains; the human groups studied include both the Historic hunting societies of the Western Plains and the prehistoric Paleoindian occupants of the Southern High Plains. Both of these occupations have attracted a considerable degree of both popular and professional attention: in particular, the popular image of native North Americans as mounted hunters who wore feathered headdresses and lived in tepees derives from the extensive documentation of the Historic tribes and the intense (and often hostile) interaction between whites and Indians on the western frontier.

From an anthropological perspective, the aboriginal occupation of the Great Plains offers an opportunity to examine many cultural–ecological relationships in a relatively simple context. Such relationships involve the ways in which human beings take advantage of patterns in the availability of the resources they need, and the Plains offer us an environment in which the bulk of these resources came from a single source: the North American bison. This assertion implies neither that other resources were not important to the aboriginal occupants of the Great Plains nor that everything native peoples on the Plains did was determined by the habits of the bison, but only that there is no doubt that the bison was the single most important resource in the region and was a resource with powerful effects on the region's inhabitants. The ability to focus on a single aspect of an environment with such far-reaching implications for the people inhabiting that environment simplifies the recognition of important relationships.

This is particularly helpful because it should be apparent in the following discussions that understanding the likely effects on human adaptations of even a single species of animal demands a level of detail in environmental analysis which is rare in either anthropology or archaeology. Environmental change and variation do not simply alter the numbers of animals in a region; they alter many of their other habits as well, often with potentially important effects on the ability of human hunters to exploit them. Similarly, changes in rainfall, on which many archaeological analyses particularly rely, are not confined simply to increases or decreases; they can include changes in the seasonal and spatial distributions and year-to-year or season-to-season regularity of precipitation, all of which affect the productivity of the region within which change is occurring.

This study is an attempt to provide, first, a summary of the information which can be used to assess at least some of the important variations in the availability of bison and other ungulates within a region, and, second, an example of how this information can be applied to examine the ecological determinants of variation in human organization in space and over time. The link suggested here between resource availability and organization depends on the size of the aggregations a society can support and the length of time it can hold them together; it is useful to note that this link does not depend on regional population size or density, as is shown later.

The utility of the study which follows can be evaluated only by the degree to which it illuminates our understanding of the ways in which people solved the problem of making a living under the various conditions with which they are and were faced. The analyses in later chapters appear to provide support of varying strength for the relevance of the theoretical relationship proposed here between environmental conditions and human organization, but are only a first step in evaluating it in depth. If this study stimulates such evaluation and helps to expand cultural–ecological analysis beyond its current limits, it will have served its purpose.

Acknowledgments

This project was originally completed for my doctoral degree at the University of California, Santa Barbara, and knowing whom to thank at the end of a project culminating nearly 13 years of undergraduate and graduate study is no small task. Although it is not true that this study, or any other piece of scientific research, is wholly an intellectual accomplishment, I have to begin this by thanking my committee: Michael Jochim, Albert Spaulding, and Michael Glassow. For 8 years these scholars guided me without trying to force me in any predetermined direction, apparently believing that I would figure out where I was going and that it would turn out to be a worthwhile place to get to. Whatever I have accomplished at UCSB was due largely to their support.

Like everyone else who has written a work of this length, I have discussed what I say in the following pages with a great many people at one time or another. Although distance has prevented me from getting as much feedback from most of them as I would have liked, comments from Bill Fawcett, Vance Holliday, Marcel Kornfeld, Dan Larson, Mary Lou Larson, John Speth, Katherine Spielmann, and Dennis Stanford have improved my thinking about many issues. Eileen Johnson first introduced me to the Paleoindian Period on the Southern High Plains as an extremely presumptuous undergraduate, for which I thank her. The site data discussed in Chapter 11 were collected for a very different project than this, and I thank Roberta Speer and Jack Hughes, of West Texas State University, Robert Campbell and Sharon Judd, of Texas Tech University, and Ann Ramadge, of the Roswell, New Mexico, BLM office, for access to the data and for help collecting and coding them. Brian Glenn spent many hours putting maps onto a computer for me, deeply endebting me to him and producing far better illustrations than I could have hoped to draft for myself. Last, he may not remember it, but Robert Bettinger set many of my thoughts in motion on this project by reminding me that the Comanche are Numic. Having thanked these people, however, I note that only I am responsible for what I say here (which Bill Fawcett in particular will be glad to hear).

The intellectual and practical assistance which these people have provided would have been for naught without the people who helped me to preserve some semblance of my humanity over the course of my graduate career, potentially a most inhuman experience. Foremost among these is Sean O'Halloran, my wife, who regularly refused to tolerate the excessively long work hours, foul moods, and monomania to which doctoral students are prone, thereby helping me to keep all of this in perspective. Paula Rudolph, assistant dean of the graduate division at UCSB, also helped me to remember what I was doing and why I was doing it. My ambition to be an archaeologist and the personal values supporting my decision to follow a path with such uncertain economic rewards come from my parents: I thank them for the ability to see a dream and the strength to follow it.

Finally, having expressed my gratitude to all of my committee members, I wish to acknowledge my special debt at the end of my student career to Albert C. Spaulding, my committee chair until he became an emeritus professor. It is an honor to have been one of his students. I have no idea how many hours I spent in Dr. Spaulding's office as he went over my logic, my statistics, and my prose, and showed me how they could all be improved. Those were not always easy hours, but where there is precision in my thought, rigor in my analysis, or clarity in my expression, it is because I spent them with him. I can think of no better way to thank him than to recognize that I am not able to think about anthropology without applying something he has taught me.

Contents

Chapter 1

Introduction

> The fates of individual nations have often been seen to hinge on chance: on the career of a remarkable man, on a code of morality perfectly suited to the times; on the happenstance of battle. It would be a dried-out person who wished to deny the roles these things have played as arbiters of our fate. But chance acts only on the stage set for it by the fundamental ecological nature of the human kind. Great leaders, fine moralities, or superior weapons of war are important only as they act on populations and ways of life that then exist. It is these populations and habits which we can understand from ecological principles.
>
> —Colinvaux 1980:13

By and large, recent anthropological analyses of human adaptations to the natural environment emphasize three topics: diet, technology, and settlement patterns. This is particularly true in archaeological research and is virtually universal in archaeological research on hunters and gatherers.

The present study grows out of a puzzling circumstance raised by this emphasis. Traditional analyses of the Paleoindian occupation of the Great Plains, and especially the Southern High Plains of West Texas and eastern New Mexico (particularly Hester 1975; Wendorf and Hester 1962), find very little variation in any of these three aspects of human behavior other than a decrease in the diversity of animals hunted after approximately 11,000 years ago and a series of changes over time in projectile point shape. Neither of these changes seems to indicate any major adaptive shifts: the post-Pleistocene dietary restriction presumably reflects the fact that most of the animals hunted before 11,000 B.P. became extinct and were thus no longer available, and it is difficult to interpret changes in projectile point style as evidence for substantial changes in other aspects of the Paleoindian way of life.

This apparent continuity is puzzling because paleoenvironmental research (i.e., Holliday *et al.* 1983; Wendorf and Hester 1975; see Chapter 9 in this volume) indicates that the Paleoindian period was marked by substantial, continuous environmental change, a pattern that is somewhat difficult to reconcile with the continuity evident in the archaeological record. The explanation for this conflict offered here is that the apparent lack of adaptive change on the Plains during the Paleoindian period reflects not a static way of life but the limits inherent in analyses that consider only the three topics noted before.

A few studies have stressed the adaptive significance of a fourth aspect of human behavior—human organization—and emphasize the adaptive problems faced by a society that attempts to make group decisions and resolve disputes without some formal means of social control (particularly Johnson 1982). Furthermore, it is apparent that many solutions to the material problems of human societies involve cooperative labor or complex regional economic interactions that are effective only in conjunction with appropriate social mechanisms (cf. Carneiro 1970; Cordell and Plog 1979:419–424).

The present study extends this approach to the apparent contrast between environmental change and cultural continuity in the Paleoindian period on the Great Plains. Essentially, it argues that adaptive changes are likely to have occurred during this period and that these changes should have involved the *organization* within which similar subsistence resources were procured, similar tools were produced and used, and similar locations were selected for occupation. It bases this argument on a detailed examination of the constraints and opportunities that the Plains environment offers to a hunting society and considers the implications of these constraints and opportunities for both the very recent and the Paleoindian occupants of the region.

1. CULTURAL–ECOLOGICAL PERSPECTIVES ON THE GREAT PLAINS

The study of aboriginal cultures on the Great Plains has a long history in American anthropology. Partly because the culture of the Plains Indians was among the very last in North America to be destroyed by the westward expansion of white settlement and partly because of the romantic impact of much of the historic Plains way of life, a substantial body of ethnographic, ethnohistoric, and archaeological research has accumulated on the inhabitants of the region. So abundant is the ethnographic documentation on the most recent Plains tribes that anthropologists have often implicitly or explicitly seen little to add to it (Eggan 1952:35–37; Kroeber 1939:76–77).

Osborn (1983:563–564) argues that this perspective has often led Plains anthropology to emphasize description and historical analysis rather than the study of adaptive process and dynamic ecological relationships. He further points out that the studies that provide the basis for our present understanding of the interaction between culture and environment on the Plains tend to examine simple correlations between general environmental zones and overall adaptive patterns.

Such studies generally view the Plains environment as relatively uniform, particularly when they discuss the hunter–gatherer groups in the region. Oliver's (1962) classic analysis of historic Plains aboriginal social organization exemplifies these studies. This analysis notes both a core of similar characteristics and a substantial degree of diversity among recent Plains hunting societies, explaining the similarities as necessary adaptations to the nature of "the" Plains environment and explicitly dismissing an ecological explanation for cultural diversity in the region. Oliver explains this diversity as the result of the different historical backgrounds of the various tribes. A similar assumption about the primary importance of historical factors in explaining the differences among Plains societies can be seen in Secoy's (1953) analysis of the effects of the introduction of the horse and the gun on Plains warfare.

Even the most powerful historical processes, though, operate in an ecological context. A society's historical background largely controls the way in which that society can exploit a given environment, but the nature of that environment also controls the success of that pattern of exploitation, and human adaptations necessarily integrate historical backgrounds and environmental realities. An adequate understanding of variation in any such adaptations demands attention to variation in both of these factors.

The general lack of anthropological attention to much of the variation in the environment of the Plains is probably due at least in part to the obvious and often monotonous uniformity in topography, flora, and fauna throughout the region, factors that do indeed set it apart from the rest of North America and create unique conditions under which humans must live. This uniformity, though, does not comprise the total environment of the Plains. The effects of climatic variability and change on the Plains grasslands are the keys to understanding the "ecological" component of cultural–ecological relationships there, and the emphasis of much of the present study is on the nature of these effects, which I will argue are critical to understanding the differences in known human adaptations on the Plains.

This point follows several important studies of the cultural ecology of the Great Plains. Wedel's (1941) analysis of the effects of fluctuations over time in precipitation and hence available moisture on the distributions of agricultural and hunting groups on the Plains is a classic example of these studies. More recently, Reher and Frison (1980; Reher 1978) have discussed the likely effects of similar fluctuations on forage conditions for bison on the short-grass Plains and the results of changes in these conditions for humans relying on these bison. Osborn (1983) has also studied climatic variation, emphasizing the importance of differences in winter severity in limiting the size of the horse herds a given region could support.

The present study is about the relationship between the nature of the great central grasslands of North America (Figure 1-1) and the ways in which human hunting and gathering cultures positioned and organized themselves in order to make a living from those grasslands. The basic problem that it examines is how the pattern of availability of needed resources in time and space affects the ways in which human beings can successfully exploit those resources. The particular emphasis here is on the organization of human adaptations rather than on what can be called techniques of adaptation, such as the use of a digging stick to collect roots or the use of a communal drive to capture any of a variety of different kinds of animals. "Organization" is used here in a general sense to refer to the network

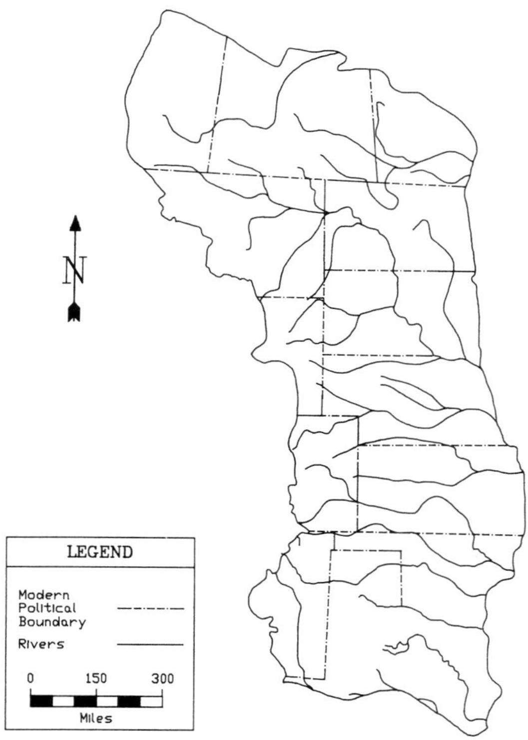

Figure 1-1. The Great Plains.

of relationships between people and groups of people within which subsistence and other activities are carried out and adaptive techniques are applied.

During most of the periods of hunter–gatherer occupation of the Great Plains, a wealth of archaeological, ethnohistoric, and ethnographic data indicate that the great bulk of food as well as materials needed for shelter, fuel, and many components of the technology derive from a single resource: the North American bison. Although other game was taken on the Plains and many plant foods were eaten, it is clear that the bison dominated both the natural fauna and the diets and much of the rest of the lives of most of the people leading a nonhorticultural way of life in the region after the end of the Pleistocene.

Although a variety of methods of hunting bison were used on the Plains (Arthur 1975; Bamforth 1985), for recent societies, the communal drive was an essential means of obtaining sufficient resources to support human populations through the year. Archaeological evidence for communal bison procurement dates back to at least the Folsom period (Dibble 1965; Frison 1978:113). The ability of the communal hunt to provide the food and other products needed to sustain human society should have varied as the availability of bison varied, and the range of environmental conditions in different parts of the Plains and the changes in these conditions in any given region over time must have contributed to this variability. The fairly dramatic differences in modern environmental conditions in different parts of the Plains place inevitably different constraints on the adaptations of any species to a given local area and thus on the adaptations of human beings preying on that species; the undoubted changes in environmental conditions over time must have had analogous effects. These effects are the major subject of this analysis.

2. A NOTE ON GEOGRAPHY

Figure 1-1 shows the western portions of the central North American grassland, the general area that is of interest here. Specifically, the study is concerned with the region from the Rocky Mountains on the west to approximately the 98th meridian south of the Nebraska/South Dakota border and to approximately the 100th meridian north of this border on the east. The northern boundary of this region is the aspen–parkland belt in central Alberta and Saskatchewan and the southern boundary is roughly defined as central Texas. A number of relatively distinct subareas can be defined within this area.

The Southern Plains refers here to the plains of Oklahoma, Texas, New Mexico, southeastern Colorado, and southwestern Kansas. The Southern High Plains, discussed in more detail in Chapters 9 through 12, includes the Texas Panhandle south of the Canadian River and the adjacent portions of New Mexico.

The Central Plains include the more northern portions of Colorado and Kansas along with all of Nebraska and southeastern Wyoming; the Northern Plains include the Dakotas and the Canadian grasslands; and, finally, the Northwestern Plains include most of Wyoming and Montana.

3. GRASSLAND RESOURCES AND HUNTER–GATHERER ADAPTATIONS

The emphasis of this study is on the relationship between the environment of the Great Plains and the adaptations of the human beings living there in the past, particularly the ways in which people distributed themselves across the landscape and the effects of this distribution on human organization. I relate these topics principally to the availability of the faunal resources in the region. This emphasis on animal foods and thus on hunting is common in Plains anthropology but seems to contradict the common generalization that temperate hunter–gatherer groups rely more on plants than animals for their food (i.e., Service 1966:10–11). Hunter–gatherer groups who rely mainly on meat are found primarily at high latitudes, particularly in the Arctic, where low temperatures and a short growing season reduce the abundance and diversity of available plants. Pointing this out, Lee (1968:42) asserts that "it seems legitimate to predict a hunting emphasis only in the Arctic."

Latitude, though, is not the only factor controlling the abundance and diversity of edible plants in an area. The cross-cultural survey on which Lee based his assertion is interesting from this perspective because it included no societies inhabiting a grassland environment. The desert-adapted groups that have dominated much of the anthropological view of hunters and gatherers, such as the Great Basin Shoshone and the !Kung San, inhabit regions in which edible plants are relatively abundant and in which animals, particularly large animals, are relatively scarce (Steward 1938:14–44; Yellen and Lee 1976:39). In contrast, most of the plants that grow in a grassland produce very little that can be eaten by human beings. However, these plants can be consumed by grazing animals, and, like the Arctic, most grasslands therefore present hunter–gatherer socieites with a relative abundance of meat and a relative scarcity of edible plants, the exact opposite of the pattern found in most other regions.

Kelly (1983:284) presents estimates of primary biomass (standing plant material) and secondary biomass (living animals) for a variety of different types of environments that highlight this difference. Table 1-1 lists these estimates and the ratio of secondary to primary productivity, measuring the relative abundance of animals and plants in each environment. (Two of Kelly's categories [lake/stream and swamp/marsh] are not considered here because they generally refer not to regional environments but to local zones within such environments.) These fig-

Table 1-1. Comparative Floral (Primary) and Faunal (Secondary) Standing Biomass (Grams/Square Meter) of Different Types of Environment[a]

Environment	Primary biomass	Secondary biomass	Secondary biomass × 10/primary biomass
Tropical rainforest	45,000	19.0	0.42
Tropical seasonal rainforest	35,000	12.0	0.34
Tropical savanna	4,000	15.0	3.75
Woodland/scrubland	6,000	5.0	0.83
Temperate grassland	1,600	7.0	4.38
Desert/semidesert	700	0.5	0.71
Temperate deciduous forest	30,000	16.0	0.53
Temperate evergreen forest	35,000	10.0	0.29
Boreal forest	20,000	5.0	0.25
Tundra	600	0.4	0.66

[a]After Kelly 1983:284.

ures indicate that the relative abundance of animals is higher in temperate grasslands and tropical savannas than in any of the other types of environments listed. As Kelly (1983:286–287) notes and as was mentioned before, the abundance of animals in grassland/savanna environments is exaggerated further for human beings because we cannot eat most of the plants that grow in them. This point particularly implies that it would be difficult to lead a nonagricultural life on the Plains without relying primarily on animals for food. The effort involved in procuring these animals must therefore have been particularly important to hunter–gatherer adaptations on the Plains, justifying the emphasis on it here.

The relative abundance of animals coupled with the scarcity of edible plants in grassland environments also leads nonagricultural societies inhabiting them to face a specific seasonal problem. Changes in the nutrition available to grazing animals at different points in the growth cycle of native range grasses generally produce a period of the year when those animals are unable to obtain sufficient nutrients to maintain their body weight (see Chapters 3 and 4). In North America, this period occurs during the winter and early spring. The result of this seasonal undernutrition is a substantial loss of body fat, which in turn affects the ability of these animals to sustain human beings.

Speth and Spielmann (1983) point out that human beings can literally starve to death on a diet dominated by extremely fat-depleted meat and that hunting societies therefore must supplement such a diet with additional fats or carbohydrates in order to obtain adequate nutrition. For several reasons (see Speth and Spielmann 1983:13–15), carbohydrates are the preferred supplement, and these are obtained primarily from plant foods. The period when these foods are most needed (winter and early spring), though, is precisely the period when they are least available naturally. The paucity of plant foods in grassland environments thus forces hunting societies in those environments into a late-winter/early-spring bottleneck where plants are essential for survival but cannot be obtained in any quantity in the natural environment.

This problem can be solved by selectively procuring only the fattest parts of the available animals during this period, by gathering and storing wild plant foods, or by trading for wild or cultivated plant foods (Speth and Spielmann 1983:18–21). Recent hunting groups on the Great Plains, for example, all made substantial efforts to obtain corn from their agricultural neighbors through trade and warfare (Ford 1972; Jablow 1951:44–46; Spielmann 1982). A major effect of this problem is that to the extent that human population density is limited by the available food supply, the population density of grassland hunters should be limited by the availability of plant rather than animal foods. Under aboriginal conditions, human populations on the Plains probably had more meat available to them than they could use.

Because agricultural products were not available to hunter–gatherer groups on the Plains before agricultural communities developed in the regions around them, smaller amounts of storable plant food were available during their seasonal bottleneck (cf. Kehoe 1978:82). This implies in turn that human population densities before agricultural products were available must have been lower than in later periods. Overall regional population density is not important to the hypothesis presented here (see Chapter 2), but this point is relevant to the discussion of Paleoindian adaptations in Chapters 10 and 11.

4. A PERSPECTIVE ON COMMUNAL BISON PROCUREMENT

The major link proposed here is between communal hunting and other aspects of human organization. Although communal hunts clearly served many purposes in Plains culture—Fawcett (1987), for example, discusses their importance in maintaining social and political relations and mating networks—this study emphasizes their importance as a source of clothing, shelter, food, and other items needed to support temporarily large social aggregations and to survive throughout the year.

Although anthropologists have described a variety of specific techniques that were used in different times and places to take large numbers of animals—herds might be run over a cliff, into a dune trap, or into a corral, for example—the organization of human effort involved in a communal hunt was essentially similar throughout the Plains. An abundance of ethnographic, ethnohistoric, and archaeological research (Arthur 1975; Davis and Wilson 1978; Ewers 1949; Frison 1973, 1974, 1978; Kehoe 1973; Reher and Frison 1980; Wheat 1972;

Wissler 1910; and many others) has described the practice of communal bison hunting on the Plains, and the following summary is drawn from these sources.

In preparation for the communal hunt, a large number of people gathered into a camp at a predetermined location under the direction of a shaman or other temporary leader. At least in recent times, these hunts occurred primarily in the fall. The locations for these gatherings were chosen because they were close to specific places that were well-suited to mass kills of bison, either because of convenient topography or because traps had been constructed there. There is evidence that individual groups of people "geared up." in preparation for these aggregations by visiting quarries and manufacturing the projectile points and butchering tools needed for the hunt. Activities in the communal encampment included both a wide range of social and ritual activities and the preparation for the kill. This preparation included such activities as sending scouts out as much as 40 to 50 miles to locate and lure in the herds and preparing the equipment needed to process the huge volumes of meat and hides to be produced.

Once the drive was accomplished, using any of a variety of relatively well-known methods of entrapment, it was necessary to kill the animals that had not died in the stampede and then to butcher the carcasses. Butchering at the immediate kill site tended to remove the major meat units for transport to a secondary processing area or to camp to be prepared and dried; this meat was often wrapped in the animal's own hide for transport. Processing the products of even a relatively small kill took several days.

Many of the details of this activity changed with the introduction of the horse and the gun, but the basic structure of the hunt remained intact. Mounted hunts were less tied to specific locations because the hunters could run the bison down on horseback instead of having to entice them into a trap of some kind. The horse also dramatically increased the size of the area over which hunters could search for their prey and reduced the difficulties of transporting the products of the kill back to camp (cf. Osborn 1983). Despite these changes, though, the aggregation of people under a temporary leader, choice of a location for this aggregation, and careful advance preparation were still necessary to a successful hunt (i.e., Walker 1982:74–94).

This brief discussion should indicate the overall predictability of the hunt and the degree of advance preparation which this predictability allowed. A detailed knowledge of bison behavior and experience in handling herds of bison apparently virtually guaranteed that Plains hunters would make a kill at a location of their choice, and it is not likely that human beings throughout the region would aggregate for a hunt year after year if the outcome of that hunt was in serious doubt. Ethnohistoric evidence (Coues 1897:576–577) documents the complete certainty expressed by at least one group of Blackfoot hunters in 1809 that they would make a kill despite repeated failures over the course of a day, confidence that was justified by their success after their white observer had left the scene of the hunt. This predictability allowed Plains hunters to organize and equip the human labor force needed to use effectively the huge volumes of meat and hides generated by large kills.

The archaeological record left by communal kills reflects the organization needed to carry them out. The typical assemblage of stone tools recovered from communal kills from all periods of prehistory in all parts of the Plains includes a fairly large number of projectile points, a few large flake knives, a stone chopper or two, a very few scrapers, and, depending on the excavation procedures used, varying numbers of small resharpening flakes (Fawcett 1986; Frison 1974, 1978; Reher and Frison 1980). The uniformity of these assemblages strongly implies that the basic organization of communal hunting changed very little over the past 10,000 years, although some of the specific tasks to which this organization may have been applied, such as the means by which the bison meat was preserved for future use, may have varied over time (Frison 1982). Stanford (1978), in fact, has uncovered apparently ritual features at the Jones–Miller Paleoindian bison kill in Colorado that are extremely similar to known ritual features in Historic kills.

A successful hunt therefore depended on being able to select in advance a specific location where the hunters were assured of killing enough bison to sustain themselves and, at least in relatively recent times, to put up stores for the winter. Reher and Frison (1980) point out that the ability to do this depends on the density of animals in a region and that communal drives are not feasible when bison populations are very low. Population density, though, is only one of the variables affecting hunting success. Herd size and mobility and the degree of dispersion of the herds throughout a region are also critical determinants of the kinds of locations where hunts can be carried out and the likelihood that enough animals can be located and captured for those hunts to be successful.

The more difficult it is to predict where herds of sufficient size to ensure survival can be taken the greater are the chances taken by people aggregating for a large hunt. If these people choose to hunt in an area that the herds suddenly abandon, for example, they face the possibility of starvation that at least some of them could have avoided by remaining dispersed in smaller social groups. This notion of predictability is the cornerstone of this study, and ungulate adaptations are considered here in the context of a hunter's ability to anticipate herd locations successfully.

In North America, the communal bison drive was confined to the Plains, but essentially identical communal hunts were conducted by societies that differed radically from those found on the Plains to procure many different kind of animals. For example, aggregation of distinct social groups under a temporary leader at a specific location, advance preparation of needed equipment, often including the construction of special trapping facilities, and division of labor by age and sex, all for the purpose of procuring a large quantity of a single species of animal, are characteristic of antelope and rabbit hunts in the Great Basin (Steward 1938:34–36) and of caribou hunts in the Arctic (Binford 1983).

Despite the homogeneity of the activity of communal hunting, it played many different roles in the lives of the different societies who employed it. Great Basin antelope drives were not necessarily carried out every year, and the products of such drives thus did not provide a reliable annual or seasonal basis for social aggregations; rather, they were one of several important resources, any one of which might be used to support any given aggregation (Steward 1938:237). In contrast, the communal bison hunt on much of the Plains appears to have been an integral part of the round of activities carried out every year, providing not only the basis for aggregations but a major source of food, hides, and other needed items. If the antelope drive failed, the Shoshone turned to pine nuts, deer, or rabbits; if the bison drive failed, the recent Plains tribes often starved.

Communal hunting can thus be viewed as a technique of food procurement that was integrated in different ways into the adaptations of the different groups who relied on it. This implies that simply knowing that a group of people practiced communal bison procurement tells us only a part of what we need to know to understand their adaptation. Even within the Plains, we can expect to see differences in the specific pattern of bison exploitation between societies in different times and places. As is discussed in the next chapter, much of this variation can be linked to the organization of these societies. The remainder of this study addresses this issue.

5. THE PRESENT STUDY

This study derives and tests a theory about the relationship between the characteristics of a region's resource base and the land-use patterns and organization of the human beings inhabiting that region. Chapter 2 presents this theory, discussing a general approach to environmental analysis, defining the characteristics of the environment to be considered here, and assessing the general effects of these characteristics on human adaptations. The emphasis of this theory on the importance to human adaptations of the natural environment does not assume that the environment operates in isolation from other forces or that other forces do not help to structure a people's way of life. The tests of this theory in later chapters potentially could have failed to support the importance of the environmental factors emphasized here. The present emphasis simply notes that people everywhere must adjust to a variety of important forces, including the natural environment, and focuses on one of these forces. Other important processes are noted in later chapters but are not the subject of this study.

The test of this theory has three major sections. The first predicts the effects of differences in the Plains environment on the adaptations of the human beings living in the region (Chapter 2) and outlines a basis for reconstructing ungulate adaptations in different times and places. It does this through a relatively detailed discussion of (1) the factors affecting patterns of forage production in a region (Chapter 3), the key to understanding bison and other ungulate adaptations and (2) the predictable responses of ungulates to these patterns (Chapter 4).

The available evidence indicates that communal bison hunting was an important part of the Plains way of life throughout most of prehistory and certainly during recent occupations of the region, and a theory linking this activity to human organization should therefore be applicable to both recent and prehistoric societies. Because archaeological data on organizational variability in prehistoric Plains hunting societies are few, the second section of this study considers the relationship between environment and organization among the most recent Plains tribes.

This second section tests the predictions outlined in Chapter 2 against the known pattern of variation in these tribes. The wealth of data on recent Plains societies provides a firm basis for examining the relationship between the range of complexity in human organization and the range of environmental conditions on the Great Plains. This section first describes the pattern of variation in climate and forage production on the modern Great Plains (Chapter 5). It then assesses the degree of difference between this pattern and that which probably existed during the eighteenth and nineteenth centuries and presents evidence on the effects of this pattern on the adaptations of the bison to different parts of the Plains (Chapter 6). Because Chapter 2 specifies a relationship between a society and the territory it inhabits, Chapter 7 summarizes the recent population movements across the Plains to indicate the territories occupied by specific groups in the recent past. Finally, Chapter 8 draws on the ethnographic record to test the relationship proposed in Chapter 2.

Having tested this relationship, the third section of this study turns to the prehistoric record in a specific section of the Plains, the Southern High Plains of West Texas and eastern New Mexico, during the Paleoindian period, between 12,000 and 8000 B.P. This section concentrates primarily on the implications of the theory in Chapter 2 for human land-use patterns because of the relative paucity of archaeological data on human organization, although it considers this latter topic as well. It emphasizes this area of the Plains because the detailed information on past environments needed to examine the issues raised here is not available elsewhere: many sites of comparable age are known on the Northwestern Plains (Frison 1978), for example, but this body of information lacks a comprehensive local paleoenvironmental framework.

This last section begins by describing the modern environment of the Southern High Plains, the nature of this environment during the Paleoindian occupation of the region, and the effects of changes in this environment on ungulate adaptations in the region (Chapter 9). Chapter 10 summarizes our understanding of the general nature of Paleoindian adaptations to the Plains as a whole and to the study area in particular, and Chapter 11 predicts the effects of the changes outlined in Chapter 9 given this understanding and tests these predictions insofar as it is possible on the basis of existing published and archival data.

Chapter 2

Resource Structure and Human Organization

The major topic addressed here is how the environment of a given region affects the complexity of human organization in that region. This chapter outlines a basis for evaluating, first, important characteristics of the resources available in a region and, second, the degree of human social complexity there. Given this outline, it then discusses a general relationship between resources and complexity, with a specific emphasis on the Great Plains.

1. ENVIRONMENTAL ANALYSIS

The goals of the environmental sections of this study are to identify for analysis distinct properties of the Plains environment that provide a basis for cross-culturally relevant generalizations about human adaptations. These goals can be clarified by examining two existing examples of environmental analysis in archaeology.

In a paper relating the organization of hunter–gatherer resource procurement to spatial and seasonal patterns of resource availability, Binford (1980) defines a continuum of organization ranging from *foragers* to *collectors*. In foraging societies, entire social groups move from resource locale to resource locale as they deplete the food available in a given local area. Such groups do not rely on stored food. In contrast, collecting societies rely on special task groups to bring distant resources to a centrally located residential base ("logistic" organization) and generally depend on stored food to get through at least part of the year.

Binford (1980:15) relates an increasing reliance on a collecting organization to increasing spatial and seasonal "incongruities" in resource distributions.

To attain the goal of identifying distinct, or independent, variables, the forager–collector continuum would be more profitably broken into two distinct responses to different properties of the natural environment. One of these is the use of special task groups to obtain food for a larger social group, a response to a heterogeneous distribution of resources across the landscape. The second of these is a reliance on food storage, a response to a heterogeneous distribution of resources over the course of the year. These two properties of the environment are not inextricably bound to one another: in principle, there is no reason why a region could not exist in which food is available everywhere in equal abundance for only part of the year, or in which food is available year-round in widely separated locations. Spatial and temporal patterns of resource availability are independent aspects of the natural environment.

The second goal, attaining cross-cultural comparability in environmental analysis, depends on describing the environment in terms of its basic structure rather than its idiosyncratic characteristics. Description of the second type is typical of much anthropological analysis. For example, Bettinger (1978:27) reconstructs two basic types of human adaptation in the prehistoric Great Basin: "a Desert Culture strategy characterized by shifting settlements and unspecialized subsistence patterns; and a Desert Village strategy characterized by fixed settlements and relatively specialized subsistence patterns." Because he notes that different specific foods were exploited by groups following each of these basic strategies and that similar kinds of foods were exploited by groups following different strategies, he argues that the differences between these two ways of life "do not reflect environmental constraints" (Bettinger 1978:27).

Identifying the species of plants and animals exploited by the people in a region, though, provides only part of the information needed to assess the relationship between environment and human adaptation. Thomas, Winterhalder, and McCrae (1979) have argued that human beings adapt to the overall spatial and temporal pattern of resource abundance and scarcity in a region and to the nature and degree of variation in this pattern rather than just to the specific species of plants and animals found there. These authors argue that many anthropological explanations are irrelevant to societies other than the one for which they were initially formulated because they rely on specific environmental characteristics, such as absolute temperatures or the species of plants or animals available, thereby obscuring structural similarities between superficially diverse regions.

Besides directing our attention to more general characteristics of a region's resources, this position particularly implies that the general structure of an environment is not adequately described by average conditions alone and that an emphasis on such conditions severely limits our understanding of cultural–ecological relationships (cf. Winterhalder 1980). No environment is completely constant from year to year, and human beings in all environments need to cope with the range of variation in resource availability caused by deviations from average conditions. This problem in understanding is exacerbated in many anthropological analyses that rely only on environmental data collected over periods of time as short as a single year. Such data can bias our comprehension of the character of both average conditions and the variability in those conditions that actually exist in a region.

To incorporate patterns of resource variation into our analyses accurately, it is necessary to obtain relatively long-term environmental data. *Long-term* in this sense refers to time periods that are long enough to reasonably depict the range of environmental variation but that are not so long that important environmental changes might occur. Given an essentially constant physiography in a specific region, climatic factors are the major determinants of the kinds, numbers, distributions, and adaptations of plants and animals in that region, and it is often possible to predict variations in resource structure from climatic data.

Variability in resource structure or *the structure of the environment,* though, are fairly imprecise phrases. "Resource structure" refers here to spatial and temporal patterns of resource availability in a particular region; the temporal portion of this definition refers both to seasonal and annual changes in availability. "Availability" refers to both the abundance or scarcity of a given resource and the accessibility of that resource to humans given their technology and the organization within which they use that technology.

This general approach to environmental analysis clearly draws heavily from evolutionary ecology, where it is particularly well developed in optimal foraging theory (Pyke *et al.* 1977; also see Jochim 1983; Pyke 1984; Winterhalder and Smith 1981). This body of theory and related aspects of ecology recognize a relatively clearly defined set of concepts describing environmental conditions (see Winterhalder 1980), although many of these concepts (such as "stability" and "diversity") refer generally to characteristics of an ecosystem as a whole rather than specifically to resource distributions and are therefore not particularly useful here. For the purposes of the present study, three general aspects of the environment that subsume the important aspects of the seasonal and spatial distribution of resources within a region are central.

The first of these, *productivity* or *abundance,* is a common component of anthropological analyses of the environment. Productivity is most simply measured by the floral and/or faunal biomass produced in a region. This total may fluctuate from season to season as animals move in and out of the region or as plants pass through their life cycles, and it may also reflect climatic variations from year to year. Thus, in a year with little rain, the total number of animals in a region may decrease as they move elsewhere to find food, and plant biomass and consequently the total amount of plant food available to human beings may also be less.

The total amount of resources available, though, may also be affected by factors other than total biomass. In an example that is particularly relevant here, Frison (1972) notes that although local bison biomass on the Great Plains was highest when the herds congregated for the rut during the summer, this was a poor time for pedestrian hunters who lacked guns and relied on communal drives to procure large numbers of animals. Bison are difficult to drive during the rut because the males, who are normally separate from the females and young, are integrated into the herds at this time. Communal hunts are therefore most likely to be successful after the rut is over. In this case, total *regional* faunal biomass reflects the amount of food available over the course of a year, but seasonal peaks in *local* biomass do not indicate the best times of year for hunting.

The second important variable is *patchiness,* which describes the degree to which resources are clumped rather than dispersed in space (Levin and Paine 1974; Wiens 1976; Winterhalder 1980). A "patchy" environment is one in which resources are distributed discontinuously across the landscape. Although such a distribution is characteristic of all environments, the degree of deviation from a continuous and homogenous pattern varies considerably. Levins and Paine (1974:2744) define a patch as "a 'hole,' a bounded, connected discontinuity in a homogenous reference background," whereas Wiens (1976:83) defines patches as areas "distinguished by discontinuities in environmental character states from their surroundings."

These definitions are deliberately general, because a specific definition of patchiness can be constructed only relative to a given organism and a given analysis of that organism. For the present analysis, one region is "patchier" than another if the first region contains needed resources in fewer or more widely separated locations than the second. Both of these criteria, though, must be evaluated relative to the ability of the human beings occupying a region to move through it. Holding environmental conditions constant, anything that increases this ability will decrease patchiness.

The concept of a patch as a spatial unit whose productivity varies from season to season and from year to year follows fairly directly from common ecological usage, but it differs substantially from the approach recently taken by Dwyer and Minnegal (1985), who attempt to define patchiness in terms of resource availability in both time and space. For present purposes, the more traditional separation of spatial and temporal patterns of resource distribution as distinct aspects of the effective environment is both simpler and analytically more tractable.

The temporal pattern of resource distribution in a region is addressed by the last important characteristic of the environment here, *predictability*. In general, a phenomenon is predictable if knowledge of some characteristic of the environment provides accurate knowledge of the state of that phenomenon. Colwell (1974) discusses predictability in reference to seasonal cycles of production, pointing out that the concept has two distinct components. The first of these is *constancy:* the availability of a resource is completely constant if it does not vary in time or space. The second is *contingency:* the availability of a resource is completely contingent if it can be predicted with certainty on the basis of the state of some other aspect of the environment. A plant that produces seeds only during one season of the year but always produces seeds during that season is perfectly contingent. Perfect predictability can derive from perfect constancy, perfect contingency, or a combination of the two. Both of these aspects of predictability can be assessed either through time, particularly in terms of seasons, or through space, in terms of location.

In this study, productivity, patchiness, and predictability must be operationalized in the specific context of the Plains environment. The basic relationship that is important here is that between the bison and human populations of the region and the perspective that guides the analysis of this relationship is that of the human hunter seeking to make a living from the bison herds, particularly through communal methods of procurement.

The overall productivity of a region can be measured by the density of its faunal population, following Reher and Frison (1980; Reher 1978). Although the bison dominated the fauna of the Western Plains throughout most of prehistory, human beings in the region also exploited other game species, particularly antelope and deer. Chapter 4 discusses evidence indicating that a single process regulates the numbers of all herbivores, and the density of all of these species should therefore follow similar patterns of increase and decrease over time. The social processes that are critical here, though, are primarily tied to human aggregations that depended on communal bison procurement, and this study therefore emphasizes this species and neglects the others.

Local patterns of productivity are described by the degree of patchiness of a resource distribution and the fluctuations in resource availability within a patch. To prey on the bison, hunters must be able to find them. This necessity provides a basis for considering the concept of patchiness. Bison and most other large ungulates live in herds of varying sizes, and the distribution of these herds determines the degree of patchiness of a given region. Conceiving a region as a mosaic of spatial units whose productivity depends on the number of animals in them, relative patchiness increases as herds of ungulates become larger and the average difference in productivity between a unit containing a herd and a unit without a herd increases. Such differences can be due either to differences in the number of areas suitable for occupation by a herd or by differences in ungulate adaptations causing differing degrees of aggregation. The degree of patchiness defined in this way is obviously likely to vary seasonally.

The size of the area a hunter or group of hunters can search for game suggests the spatial scale at which patches should be conceived for this study, although no attempt is made here to quantify patchiness rigorously. Assiniboine runners searching for herds for communal drives occasionally went as far as 40 to 50 miles from camp (Arthur 1975), but more common search distances were less, possibly on the order of 5 to 10 miles. Mounted hunters could obviously cover much greater distances than this, and the relevant patch size therefore increased after the introduction of the horse.

Fluctuations in the resources available within a patch fall under the concept of predictability. As was noted in the preceding chapter, predictability is defined in terms of a hunter's ability to find a herd. If knowledge of some environmental condition enhanced this ability, the herds in a region could be said to be a predictable resource. Such knowledge could derive from information about a regular seasonal pattern of ungulate migrations or of the spatial distribution of resources regularly used by animals: if only one source of water existed in a region, for example, the area around that source would probably have been a good place to hunt. A third basis for predicting herd locations would be a detailed knowledge of short-term, local environmental conditions and of repetitive ungulate responses to them.

The two distinct components of predictability, though, must be evaluated separately. When the productivity of the patches in a region does not change, resource constancy is high; when it changes often or dramatically, resource constancy is low. Bison and other ungulates are obviously mobile resources that do not remain perpetually in one place. Increases in the degree of mobility of such animals (that is, in the frequency of movements, the distance of movements, or the speed of movements) therefore decrease the constancy of a region's effective environment. In addition, factors that cause great changes in herd size from season to season or from year to year reduce the degree of constancy in resource availability: a region is more constant when herd sizes remain relatively similar over time.

Contingency refers to the degree to which future conditions of resource availability can be predicted from a knowledge of present conditions. The greater the regularity of ungulate movements in a region within a year and the less the variation in ungulate numbers and movements between years, the higher is the degree of contingency of the resource base in that region. When the numbers of animals present or their patterns of aggregation and movement fluctuate erratically within a season or from year to year, contingency is low.

Productivity, patchiness, and predictability can then be assessed for this study by considering the numbers of animals in a region, the size of the herds they form, the degree to which herd sizes change over the course of the year and from year to year, the frequency, distance, and speed of herd movements, and the regularity of those movements within a season and from year to year. Chapters 3 and 4 show that these factors are all strongly related to climatically determined patterns of forage production, and Chapter 5 considers variation within them on the Great Plains.

Ecologists generally explicitly quantify variables such as these, but such an effort is beyond the scope of this study. Accurately measuring the variables defined here is a complex problem even for modern environments (for ungulate biomass, see Grobler and Jones 1980; Meissner 1982; Mentis 1977; Mentis and Duke 1976; for patchiness, see Wiens 1976; Winterhalder 1980; for predictability, see Colwell 1974); these problems are obviously dramatically greater for prehistoric environments. The goal of the environmental sections of this study is to take the first step in assessing ecological conditions on the Plains by identifying the nature and direction of differences in resource structure in different parts of the region and over time in one specific part of the region—the Southern High Plains. Quantifying these differences in an absolute sense is a second step for future research.

2. ORGANIZATIONAL COMPLEXITY

This study is particularly concerned with human *organizational* responses to resource structure. The specific aspect of these responses that is important here is the degree of social complexity present in a society. Although the study of complexity has been important in anthropology for some time, traditional approaches to this topic are somewhat difficult to apply to human societies on the Plains.

First, as McGuire (1983) points out, these approaches lump together two very different concepts into a single variable; McGuire refers to these two concepts as *heterogeneity* and *inequality*. Heterogeneity refers to the number of social categories in a society and the distribution of individuals across these categories: heterogeneity is high when there are many categories containing comparable numbers of people and low when there are either few categories into which individuals can be placed or when most individuals are in a small subset of all of the possible categories. Inequality refers to differences in access to "material and social resources, such as wealth and power" (McGuire 1983:101–102). Inequality is high when a few individuals have access to substantially more such resources than most of the other people in their society, and it is low when resources are equally available to all. As McGuire's analysis shows, these two variables need not always vary together.

Furthermore, when social complexity is discussed by anthropologists, it is nearly always in reference to societies that are obviously "complex" in some intuitively sensible way. The goal of such discussions is often to explain how a particular degree of complexity, generally conceived as a stage of development such as a chiefdom or a state (Service 1962), came to be. When complexity is broken down into its two constituent parts, heterogeneity and inequality, it is clear that such "stages" are essentially arbitrary divisions of a continuum from the simplest hunting and gathering societies to modern industrial states and that each stage subsumes a wide range of variation. It is therefore important to examine differences in complexity within as well as between stages. In addition, changes in each of the two components of complexity represent distinct kinds of human responses to distinct kinds of situations. Lumping them together obscures the actual processes affecting social complexity and makes it difficult to distinguish the range of diversity among societies in which heterogeneity and inequality are not both highly developed.

Among the known hunting societies on the Plains and among many (although by no means all) other hunting and gathering societies, differences in complexity were primarily linked to differences in heterogeneity. Differences in inequality based on the size of personal horse herds are known to have developed in at least some Plains tribes during the nineteenth century, and it is clear that greater wealth enhanced an individual's ability to become a chief (e.g., Ewers 1955:240–255). However, it is also clear that individuals held a chief's position only as long as their personal characteristics qualified them for it and that the actual process by which tribal decisions were made relied heavily on concensus (especially see Grinnell 1962a:338–342).

Complexity for the purposes of this study therefore refers primarily to heterogeneity as defined before. This concept of complexity is similar to that found in ecological research: Colinvaux (1973:237) defines a complex system as "one with many parts and interactions between those parts," and Odum (1971) relates human social complexity to the number of different occupations in a society. Definitions such as these imply a continuum of increasing complexity from minimally differentiated organizations to organizations with many parts. Such a continuum underlies much of this study, and differences in social complexity are conceived here as differences in the number of recognized social categories and groups within society.

The relationship between heterogeneity and resource structure can be approached by considering what it is that increasingly complex social organizations do. One key to this question seems to lie in the relationship noted by many anthropologists between population size and organizational complexity. Although they have measured complexity in many different ways, these authors (e.g., Carneiro 1967; Ember 1963; Naroll 1956) have uniformly found that there is a strong, quantifiable relationship between greater population size, particularly local population size, and more complex social organization. This relationship is commonly explained by the need for more effective means of social control when a larger number of people are in continuous face-to-face contact with one another and by the difficulty of reaching decisions by consensus with large numbers of people. The correlation between population and complexity, however, is not perfect, and there is considerable variation in the degree of complexity seen among populations of comparable size (Johnson 1982:390–391). The explanation just noted for the demonstrated relationship between population and complexity depends on *local* population size and not directly on *regional* population size or density, a critical point for the discussion and analyses that follow.

At the relatively simple level of organization present among known Plains hunter–gatherers, mechanisms for social control and decision making tend to be flexible and seasonally effective, following seasonal changes in the size of human aggregations (see Lowie 1954:125–126). Johnson (1982) has explored situations such as this in some detail. Lacking a centralized formal authority structure, these simpler societies generally arrive at decisions by consensus, which requires face-to-face personal interaction. As decisions involve larger and larger numbers of people, such interactions become more and more difficult to maintain, because there are limits on the number of people who can effectively work together in this way that may derive from inherent limits on human information-processing abilities. Furthermore, disputes between individuals become more and more frequent as group size increases. Johnson's (1982:396–402) analysis of data on !Kung San group sizes and dispute frequencies indicates that this is caused by increases in the size of local residence units rather than by overall population density in a region.

Johnson (1982:396–407) suggests that, among societies that lack formal social hierarchies, these problems are often solved by developing "sequential hierarchies." In a society organized in this way, decisions by consensus can be reached by the smallest organizational units of the society, and these decisions can be considered by representatives of each of these units, progressing eventually through larger and larger components of the social group. For example, "if consensus were achieved first within nuclear families, then within extended families, a group decision would only require consensus among extended families" (Johnson 1982:403). He argues that the number of individuals involved in actually making decisions remains relatively constant but that, as group size increases, these individuals come to represent larger and larger numbers of people. Thus, although smaller wet-season !Kung camps are organized by nuclear families, larger dry-season camps are organized by extended families: as group size increases, the size of the units on which the group's organization is based increases, but the number of these units remains essentially the same.

Increasing the complexity of a sequential hierarchy is one way of increasing social heterogeneity (cf. McGuire 1983:107–108). However, sequential hierarchies cannot be expanded indefinitely: as the number of steps required to reach a decision increases, so does the time required and the chances that consensus will not be achieved. As group size continues to increase, then, an alternative solution—developing a "simultaneous" hierarchy, or a nonegalitarian social order in which decisions are made by an elite—becomes more likely; the alternative is group fissioning (Johnson 1982:407–416). Such a development corresponds to an increase in inequality as it is defined here. However, this process is unlikely to be important in the context considered here, where the alternative solution, group fissioning, is an inherent seasonal component of the adaptation being examined.

This therefore implies that among societies with a seasonal pattern of aggregation and dispersion, increasing local population size should lead to increasing heterogeneity in order to maintain social control and to reach group-level decisions effectively. The structure of the resources in the environment that a society inhabits is related to this process by examining the degree to which it encourages or inhibits social aggregations and provides the means of supporting more heterogeneous social structures. The development of a regular pattern of human aggregation requires external conditions favoring aggregation, a mechanism leading individual families to aggregate and a sufficiently productive local resource base to support large groups of people while they are together.

As heterogeneity increases with larger and/or more permanent aggregations, additional requirements appear. First, for a distinct subgrouping within a society to maintain its identity, its members must have contact with one another on a regular or reasonably regular basis. This is probably particularly true when subgroups begin to develop that are not based strictly on kinship ties. Second, as Johnson (1982:405–407) notes, large aggregations in relatively simple societies are often tied to major ceremonies. Such ceremonies generally require specific ritual paraphernalia and relatively great cooperative effort to construct special ritual facilities. The manufacture and maintenance of these sorts of items and facilities requires that sufficient time and materials be available before the time when the ceremony is to occur. As group size increases, membership in subgroups is also often signaled by specific items of material culture (Conkey 1978; Hodder 1979), and production of such items requires investments of time and materials. As social groups become larger and more heterogeneous, then, it is necessary to maintain fairly regular associations of specific individuals and to ensure adequate time to produce and maintain artifacts and facilities that symbolize and support the pattern of social organization and interaction.

3. RESOURCE STRUCTURE AND HUMAN ORGANIZATION

Both the requirements of developing and physically maintaining human aggregations and the requirements of supporting more complex social organizations can be linked directly to the nature of the available resource base as well as to other factors. In nomadic societies, forming relatively large groups confers a number of advantages on individual human beings (cf. Fawcett 1987). Many resources are harvested more effectively by larger than by smaller groups of people (Frison 1978; Smith 1981). Large groups drawn from a large area also bring information about environmental and other conditions over a large region together in one place, allowing smaller social units to plan their future movements (Conkey 1980; Thomas 1972), and they also provide a larger pool of potential marriage partners, increasing the likelihood of finding a spouse (Wobst 1974).

These factors suggest that, under most conditions, aggregation is likely to confer some kind of benefit on the individuals aggregating and therefore should be favored (Jochim 1976:68–69). Communal hunting can thus be viewed not only as a direct response to the labor requirements of controlling and processing large numbers of animals but also as a mechanism that permits human aggregations to occur and provides the means for sustaining them, as Fawcett (1987) argues. The effectiveness of this mechanism, however, depends on the reliability and productivity of the hunt. Limits on the ability of a social group to come together in one place to hunt or for some other purpose can therefore be linked to environmental conditions in most cases: it must be possible either to produce a sufficient amount of food in a local area to support an aggregation or to store up sufficient supplies in advance to last through the period of aggregation.

It is important to distinguish here between overall human population densities and the degree of population aggregation. At least at lower densities, it is possible in principle for regional populations to increase substantially without triggering any organizational change, so long as there are no changes in the frequency, size, or duration of social aggregations. Schalk (1979:69) does not make this distinction clearly, arguing that social complexity on the Northwest Coast was not linked to population density, which he measures as human beings per square mile. In fact, Schalk's data (1979:64–66) show a clear north-to-south increase in mean winter village size, or *local* population density, paralleling a similar gradient in social complexity (Schalk 1979:67–68; Suttles 1968:64–65).

Although it is likely that some degree of aggregation is always preferred by human beings, the timing and size of aggregations are extremely variable. Jochim (1976:65–79) discusses this issue and summarizes a variety of ethnographic evidence that social group size is linked closely to resource availability, suggesting that:

> the greater degree of spatial concentration of resources, the greater the yield of a single site catchment as defined by distance-minimization considerations, and therefore, the larger can be the coresiding group. Conversely, the more dispersed the resources, the smaller the catchment yield, and thus the smaller the group supportable. (Jochim 1976:66)

Specifically, Jochim (1976:23, 72–74) argues that the size of a social aggregation should be proportional to a measure of resource availability equal to:

$$wnad/m$$

where *w* equals the weight of an individual member of a prey species, *n* equals its nonfood yield, *a* equals the size of the aggregation in which it is found, *d* equals its overall density in the environment, and *m* equals its degree of mobility.

All of these variables are demonstrably important to hunter–gatherer decision making (Jochim 1976:22–24). This measure summarizes the regional abundance, concentration, and ease of locating and acquiring a given resource; this hypothesis thus predicts that aggregations should be larger when they are supported by resources that are presented in large units, satisfy needs in addition to that for food, occur in large aggregations, are abundant in the environment, and are not very mobile. This prediction is supported by data for the Round Lake Ojibwa (Jochim 1976:73–74).

Maintaining the social structure needed to integrate larger groups of people depends also on other environmental characteristics. Human beings exist in a world that contains a finite amount of resources and provides them with a finite amount of time in which to harvest those resources in order to satisfy their needs. Time limitations on human activities are particularly important (Smith 1979; Torrance 1983): because humans cannot control the seasonal patterns of resource availability that determine the time when many activities can be carried out, it is generally necessary to schedule activities so that the full range of human needs can be satisfied.

Production and maintenance of social and ritual regalia or facilities for seasonal aggregations are generally activities that must be so scheduled. As providing for the biological needs of a human population takes up more and more time, less and less time is available for other activities. Furthermore, activities such as this often cannot be scheduled at will but must be scheduled during a period that is appropriate for the timing and location of the anticipated aggregation. To prepare reliably for aggregations, then, it is necessary to know that sufficient free time will be available at a given time of the year.

The location and timing of seasonal human aggregations such as those found among most of the recent Plains tribes also tend to be determined several months in advance of the actual time they occur. This requires that a suitable location be specified without knowing exactly what the conditions at that location will be like. Among mobile hunting people, the alternative to this is to collect the social groups that happen to be nearby when a suitable time and place for a gathering is found. To maintain a regular association of specific individuals and to ensure that the individuals and equipment needed for ceremonies at a gathering are actually present, the former system is preferable. However, it is feasible only when the characteristics of an aggregation site can be known in advance with some certainty.

This requirement relates closely to the concept of resource predictability discussed before. Both regularity of attendance and the likelihood that the personnel and regalia needed for ceremonies will be present will obviously decline as the chances increase that the resources at a given aggregation location will be inadequate, and the effort needed to support a greater degree of social complexity will therefore be more difficult to sustain. *Ad hoc* aggregations, probably smaller with shifting memberships and relying less on major ceremonies to integrate the people at the gathering, should then become more common in environments that are unpredictable in the sense in which this term is used here.

Ecologists have discussed several aspects of this relationship more rigorously. However, there are a number of assumptions underlying ecological theory that make it difficult to apply to human beings in some circumstances. First, much of this theory, and especially those portions of ecological theory with which anthropologists are most familiar, addresses foraging behavior and assumes that this behavior can be predicted only on the basis of caloric or other nutritional return for time or energy expended. Second, this theory generally relies on computing resource "costs" in terms of nutritional return on time or energy expended locating, capturing, and processing resources. Once such costs are computed, this theory assumes that they are immutable. Last, this theory assumes that the organisms whose behavior it predicts are capable of processing the information needed to evaluate such costs and make the decisions it expects them to make.

There are good reasons to question all of these assumptions. Human beings, at least, obtain needed resources other than food while foraging; material for clothing, shelter, and tool and other artifact manufacture are the most obvious of these other resources. The need for materials such as these must also enter into foraging decisions but is not considered in most ecological models attempting to predict these decisions. Perhaps less obviously, resource costs are not fixed for human beings. The costs of locating, capturing, and processing a resource depend on the technology and labor organization available to a society and can change rapidly and profoundly when useful innovations in either of these become available through invention or diffusion. Finally, as Jochim (1983) discusses, it is quite likely that human beings make decisions about resource procurement on the basis of partial information or by considering simple proxy measures of resource return, such as animal size, that may or may not accurately reflect nutritional considerations. These considerations imply that ecological theory must be used cautiously in anthropology and that it is more likely to be useful at a fairly general level of analysis than in making extremely specific predictions of human behavior.

Bearing these possible problems in mind, two ecological models are particularly relevant here. First, Horn (1968; also see Wilmsen 1973; Winterhalder 1981:30–32) has considered how foragers should locate themselves within a region given several different resource distributions, assuming that an optimal locational strategy minimizes the distance that must be traveled to obtain food. When resources are dispersed and immobile, this distance is minimized when foragers are dispersed throughout a region; when resources are mobile and clumped, this distance is minimized when foragers are aggregated in a central location. Heffly (1981) has found that the general predictions of this model are consistent with the adaptations of Northern Athapaskan foragers.

However, there are limits on this model's applicability, because it does not address foraging behavior under all possible environmental conditions: there are predictable sets of conditions that it does not consider. This problem arises because mobility and degree of dispersion are distinct characteristics of a given resource. Although the model predicts foraging behavior for dispersed/immobile and clumped/mobile resources, it neglects behavior for dispersed/mobile and clumped/immobile resources. It is therefore not useful in all cases.

Bryant (1973) presents a more complete analysis of population responses to a patchy environment. This analysis attempts to predict a population's *grain response* to varying environmental conditions. This response is *fine-grained* if population is dispersed across habitats of unequal quality in proportion to the productivity of those habitats (if population "tracks" habitat quality); it is *coarse-grained* if population is concentrated in the most productive habitat and lower quality habitats are not used.

To examine the determinants of a population's grain response, Bryant constructed a mathematical model that considers the differences in mean quality within a set of habitats, the variance in this quality within a habitat, and the autocorrelation of successive conditions within a habitat. The degree of difference in quality between habitats corresponds to patchiness as it is defined here. Variance in habitat productivity corresponds to constancy: invariant environments are constant. Finally, autocorrelation in Bryant's study measures the degree to which future conditions can be predicted on the basis of present conditions, a definition that essentially parallels the concept of contingency defined here. The remainder of this discussion substitutes these terms for Bryant's terminology.

Bryant's analysis indicates that in a patchy environment, a coarse-grained, or aggregated, population distribution is optimal when constancy is high regardless of the degree of contingency in the environment. As constancy decreases, the distribution of the population should be more fine-grained, or dispersed, particularly when contingency is high. These conclusions, though, do not take the beneficial effects of being in an aggregation into account; they consider only the effects of patch productivity.

As is discussed before, human beings should prefer to aggregate for reasons that have nothing to do with the environmental characteristics of any specific aggregation site. When contingency is high, future conditions can be predicted accurately on the basis of present conditions, making it relatively easy to plan for an aggregation in advance. In a patchy environment, human population responses should therefore be coarse-grained when contingency is high regardless

Table 2-1. Population Responses to Environmental Characteristics[a]

Patchiness high

		Contingency High	Contingency Low
Constancy	High	Aggregate	Aggregate
	Low	Aggregate	Disperse

Patchiness low

		Contingency High	Contingency Low
Constancy	High	Disperse	Disperse
	Low	Disperse	Disperse

[a]Modifed from Bryant 1973.

of the degree of constancy in the environment. Table 2-1 summarizes Bryant's conclusions, modified to take this into account.

Phrased in terms of the environmental variables discussed in the first section of this chapter, the relationship proposed here among resource structure, human aggregation, and organizational complexity can be stated relatively explicitly. For the present analysis, the critical properties of the environment are the total ungulate population in a region, defining regional productivity; the size of the herds in a region, defining local productivity and contributing to the degree of patchiness; the degree of dispersion of the herds, also contributing to patchiness; the overall degree of herd mobility and the regularity of the pattern of herd movements from year to year, defining contingency; and the degree of variation in regional population size and in herd size, dispersion, and mobility from season to season and from year to year, defining constancy.

When herds are small and dispersed more or less evenly across the landscape, a fine-grained distribution of the human population preying on them is expected. When this condition persists throughout the entire year, human aggregations should be small and infrequent, and social heterogeneity should be low. Substantial increases in the patchiness of herd distributions during one season of the year should be accompanied by increasingly aggregated human populations and hence greater heterogeneity, particularly when herd dispersion and mobility are likely to be high in the following season(s) and substantial numbers of animals will therefore be relatively difficult to locate. Smaller seasonal changes in herd dispersion relative to the overall degree of fluctuation in herd sizes and mobility will not have this effect. When conditions favoring aggregation persist for more of the year, people should remain in larger groups for longer periods of time, and heterogeneity should increase with the need for more effective social control.

The relationship between bison ecology and human organizational complexity proposed here thus predicts that less complex societies should be favored when the herds are small, dispersed, and move often or over long distances in irregular patterns across the landscape and should also be favored when these patterns vary substantially from year to year. Increases for all or part of the year in bison population size, herd size, and the regularity of herd movements, or decreases in the degree of herd mobility, should favor more complex societies.

Chapter 3

Grassland Ecology

This chapter discusses the factors affecting the annual and seasonal patterns of forage production in a grassland, anticipating the discussion in the following chapter showing the importance of these patterns to ungulate adaptations. The major subjects of this chapter are the effects of climate, soils, and grazing on the amount and quality of forage a grassland can produce. Attempts to specify mathematically the biological, chemical, and environmental variables determining grassland productivity clearly indicate the enormous complexity of the relationships among them (Innis 1977). For analyses at the relatively large scale of this analysis, though, it is the general effects of these variables on the grassland ecosystem that are important, and these effects are relatively well known.

This discussion focuses on two aspects of grassland productivity: the amount of forage produced in a region and the nutritional quality of that forage. Although grasses provide many nutrients for grazing animals, this chapter discusses only three of them that are particularly emphasized in the literature: carbohydrates, protein, and phosphorus, all essential for bison and other ungulates. Besides these specific nutrients, the digestibility of the forage, usually measured as the percentage of the plant material that can be digested by a grazing animal, is important here. Both forage production and nutritional value are affected by the maturity (phenological stage) of the plant and by external factors, particularly climate, soils, and the intensity of grazing.

1. TYPES OF GRASSES AND TYPES OF GRASSLANDS

General types of grasslands are distinguished by the various types of grasses found in them. Grasses can be grouped in several different ways (see Weaver

1954:8–15), depending on the problem at issue, but the most common distinctions are based on height, with three classes recognized: tall grasses that grow to heights of 5 to 8 feet, midgrasses that grow to heights of 2 to 4 feet, and short grasses that grow to heights of 0.5 to 1.5 feet. On the Great Plains, common tall grasses include big bluestem (*Andropogon gerardi*) and switchgrass (*Panicum virgatum*); common midgrasses include little bluestem (*Andropogon scoparius*), western wheatgrass (*Agropyron smithii*), sideoats grama (*Bouteloua curtipendula*), bluebunch wheatgrass (*Agropyron spicatum*), and needle and thread (*Stipa comata*); and common short grasses include blue grama (*Bouteloua gracilis*), hairy grama (*Bouteloua hirsuta*), buffalograss (*Buchloes dactyloides*), and, particularly in the south, galleta (*Hilaria jamesii*) and black grama (*Bouteloua eriopoda*). The presence and relative abundance of these three types of grasses depend on available soil moisture, with the tall grasses dominant where moisture is most abundant and the short grasses dominant where it is least abundant.

A second major classification of grasses is based on their phenology, or pattern of growth. This classification refers to the time of the year when grasses begin to grow, and recognizes "cool-season" species that begin to grow in the spring, complete their growth cycle by summer, and sometimes have a second peak of growth in the fall, and "warm-season" species that complete a single cycle of growth during the summer (Weaver 1954:9–10). Several studies (Boutton *et al.* 1980; Kemp and Williams 1980; Monson and Williams 1982) have demonstrated that temperature is the critical determinant of the relative abundance of cool- and warm-season species in a region. During cooler portions of the growing season, the bulk of grass production is in cool-season species: during warmer parts of the season, this role is taken by the warm-season species (also see Ode *et al.* 1980; Sims and Singh 1978a:557). Furthermore, in regions with overall higher annual temperatures, a larger proportion of the total grassland community is comprised of warm-season species than in regions with cooler overall temperatures (Terri and Stowe 1976).

Although this simple dichotomy is very useful, it is somewhat oversimplified. Dickinson and Dodd (1976) found a continuum of growth starts from early April through mid-October among the 34 range plant species they studied. Their data indicate that four types of plants can be defined: those that flower only once during the early part of the growing season, those that flower both early and late in the season, those that flower only once during the middle of the season with the specific time of flowering determined by local moisture conditions, and those that flower only once late in the season. Together with the studies noted before, these results indicate that grassland communities can be placed on a continuum from those in cooler regions where relatively substantial amounts of plant growth occur from spring until midfall to those in warmer regions where such growth is largely confined to the summer months. The pattern of forage production over the course of the year is also more erratic in warmer areas than in cooler areas because of the greater proportions of warm-season species such areas contain and the closer relation of growth in such species to local rainfall (Sims and Singh 1978a:557, 562).

There is also some evidence indicating that cool- and warm-season species differ not only in seasonal growth pattern but also in nutritional value (Caswell *et al.* 1973; Coppock *et al.* 1983a). First, physiological differences between these two groups of plants appear to lead warm-season species to contain lower protein concentrations than cool-season species (Brown 1978). In addition, because warm-season species tend to store more protein in more durable plant parts than do cool-season species, they are also less digestible (Akin and Burdick 1977). This last difference appears to be greatest in the fall, when one study found cool-season grasses showing 15% to 20% greater digestibility than warm season species (Coppock *et al.* 1983a).

2. STAGES OF GROWTH AND NUTRITIONAL QUALITY

The nutritional value of native range plants changes as they progress through their growth cycle. A grass's nutritional value is highest during the early portions of the growing season and declines substantially as the plant matures, generally dropping during its dormant winter or dry-season stage below the point where it provides sufficient nutrition for most large grazing animals to maintain their body weight (Sinclair 1975, 1977; Speth 1983). This pattern has been demonstrated for protein content (Cogswell and Kamstra 1976; Everitt *et al.*, 1982; Goetz 1975; Hart *et al.* 1983; Rauzi 1975; Rauzi *et al.* 1969; Sampson 1952; Willard and Schuster 1973; Weinmann 1955), phosphorus content (Everitt *et al.* 1982; Rauzi *et al.* 1969; Sampson 1952; Weinmann 1955), and digestibility (Cogswell and Kamstra 1976; Hart *et al.* 1983; Sampson 1952).

However, this decline is not necessarily constant over the course of the growing season. Growth in Plains grasses, particularly warm-season grasses (Dickinson and Dodd 1976; Sims and Singh 1978a:557, 562) tends to be a direct response to even small amounts of precipitation (Rauzi and Dobrenz 1970; Sala and Lauenroth 1982; Sims *et al.* 1978) and therefore occurs not as a continuous process but in pulses following rainfall. Protein, phosphorus, and digestibility all increase as a direct response to local rainfall (Everitt *et al.* 1982; Hart *et al.* 1983; Rauzi *et al.* 1969; Willard and Schuster 1973). Where growing season rainfall is relatively continuous or cool-season species are relatively abundant because of relatively low temperatures, seasonal changes in the nutritional value of forage in a region will be relatively regular; where growing season rainfall is more episodic and higher temperatures produce grasslands dominated by warm-season species, seasonal changes in nutritional value will be more erratic, with grasses being more nutritious following precipitation.

Although this overall pattern is characteristic of all Plains grasses, as was noted earlier, there are significant differences between species in the timing of the various stages of growth. Because they start to grow earlier in the year, cool-season grasses will have their highest nutritional value before warm-season species, and a grassland containing a mix of warm- and cool-season species will therefore have more grass of greater nutritional quality available through more of the year than a grassland dominated mainly by warm-season species. Highly nutritious forage may be found well into the fall if the appropriate species are present in the area.

3. CLIMATE AND GRASSLAND PRODUCTIVITY

Precipitation and temperature are the two critical aspects of the climate for the purposes of this discussion. Of these two, precipitation is the more important. Some of the effects of precipitation patterns were just noted, particularly the immediate increase in the growth rate of many native range grasses in direct response to local moisture conditions. Like other grasslands, the North American Plains are largely a water-limited area (Brown 1977). Thornthwaite (1941), for example, notes that agriculture in the region has historically been controlled by drought, and Wedel (1941) argues that long-term fluctuations in precipitation determined the western limits of aboriginal horticulture, with purely hunting cultures exploiting the more arid Western Plains. The sensitivity of the Plains grasses to changes in available moisture has been extensively documented during the twentieth century and is critical to understanding prehistoric adaptations in the region. Variability in available moisture affects grasslands by altering (1) the density of plants in a given area, (2) the relative proportions of species present, and (3) the physiology of the plants present. The following summary of the first two of these is largely drawn from Coupland's (1958) review.

Aridity affects grasses by creating "water stress," which occurs "when the rate of water loss by transpiration exceeds the rate at which it is replaced by absorption" (Brown 1977:107). Because plants absorb water from the soil through their roots, drought particularly affects grasslands by reducing soil moisture. Under drought conditions, this reduced moisture can result in a rapid and substantial decrease in forage yield. Coupland (1958:288), for example, notes a decrease in forage yield in Montana during during the 1930s Dustbowl from 1586 pounds per acre in 1927 to 222 pounds per acre in 1934, one result of which was a parallel drastic reduction in the carrying capacity of the affected region for herbivores such as cattle and bison. Reduction in forage is caused by a decrease in the height of individual plants, by the wilting and death of less drought-resistant species, and by a reduced vigor in more durable species. In a prolonged drought, this results in a dramatic decrease in ground coverage: during the great drought of the 1930s, much of the Great Plains saw a decrease in basal coverage in some areas from 89% to 22% in only 5 years, and many areas had a coverage of only 5%. Uncontrolled grazing also accelerates reductions in basal coverage during drought.

The effects of increased moisture on the Plains vegetation are essentially opposite to those of drought. Such increases are followed by increased forage yield and basal coverage and, if thev are pronounced enough, the invasion of midgrass species into short-grass communities and of less drought-resistant mid- and tall grasses into mixed-grass communities (Coupland 1959; Newbauer *et al.* 1980). In ungrazed communities, forage production increases with increasing growing season precipitation up to 500 mm/year and then levels off, but in grazed communities, this increase is linear up to at least 1000 mm/year (Sims and Singh 1978b:579).

To summarize, the overall effect of increased precipitation on the Plains grasslands is to increase their productivity, at least as productivity is measured by forage yield. Conversely, decreased precipitation decreases forage yield. Some research (Sneva 1976) suggests that annual grassland productivity is particularly related to spring precipitation. The effects of variations in precipitation on plant physiology that cause these changes are extremely complex (see Hsiao 1973; Innis 1977), and a comprehensive overview of them is beyond the scope of this discussion. Several key aspects of the relationship between plant physiology and aridity, though, are important here.

Less severe water stress than that which occurred during the 1930s affects grassland productivity primarily by altering the nutritional value of the available grass rather than by dramatically changing the forage yield. Although the data on many of the nutritional effects of water stress are incomplete, certain patterns are evident. This discussion will concentrate on how available moisture affects the concentration within a plant of carbohydrates, protein, and phosphorus content, the three nutrients emphasized here.

Stored carbohydrates form the major source of energy available to ruminants from plants. Carbohydrates are produced during photosynthesis and subsequently stored to be used when the plant is growing. They are distributed from storage areas to different portions of the plant as they are needed (Bokhari 1978a; Moser 1977; Trlica 1977; White 1973). Unfortunately, the effects of water stress on the concentration of carbohydrates present in a plant are poorly understood: studies have found both that water deficiency sometimes increases and sometimes decreases the total reserve present in a plant, possibly depending on the point in the growth cycle of the plant when it occurs (White 1973). However, because stored carbohydrates are distributed within a plant as they are needed, the effects of water stress for a grazing animal are relatively clear. By reducing the growth above the ground, water stress leads grasses to store carbohydrates in their stems and roots instead of in their leaves (Bohkari 1978a; Moser 1977), making them essentially inaccessible to grazers. Furthermore, when the growing season is short, stress conditions during one year can deplete a plant's carbohydrate reserves, lowering its productivity in subsequent years (Trlica 1977).

Variation in available moisture affects the protein and phosphorus contents of Plains grasses by altering the availability of nutrients from the soil. Nutrients from the soil are obtained by plants in solution in soil water through the roots. One major effect of extended drought on the Plains is severe depletion of soil moisture: Tomanek (1959), for example, notes a complete absence of moisture to a depth of 5 feet in the soil in part of Kansas in 1939 following the Great Drought. Such conditions drastically reduce the amount of nutrients that are available to plants, particularly nitrogen, probably the single most important nutrient determining grassland productivity (Cole *et al.* 1978; Reuss and Innis 1978). Low nitrogen content in a plant not only reduces overall forage yield but also lowers the rate at which phosphorus is absorbed (Cole *et al.* 1978). Arid conditions thus reduce both the nitrogen and phosphorus content of grasses, whereas less arid conditions increase them.

Less effective nitrogen metabolism due to water stress also reduces the rate of protein synthesis in plants. A variety of evidence indicates that even mild increases in aridity can have powerful effects on this rate, although this process can be reversed quickly when moister conditions develop (Hsiao 1973). This implies that plants that experience more or less constant moisture deficiencies will tend to have lower protein contents than plants that do not experience such deficiencies (Crowder and Chheda 1982:351).

Temperature changes also affect the forage quality of Plains grasses. The various parts of individual grass plants can be conveniently divided into the cell contents and the cell walls. The cell contents are digestible by ruminants such as bison and contain the bulk of the nutrients they obtain from forage, but the cell walls by and large are not digestible. As temperatures increase, so does the rate at which plants mature, resulting in greater perceptages of undigestible crude fiber in their edible parts than are found in plants grown at lower temperatures. As was discussed before, the overall nutritional value of grasses, including crude protein and mineral content, also decreases as they mature, and more rapid maturation rates thus decrease the length of the period during which the plants have a high nutritional value.

It is particularly important to note that these precipitation- and temperature-dependent physiological processes begin long before there is *any* macroscopic change in plant appearance, ground coverage, or species composition (Brown 1977; Hsiao 1973). The effects summarized before from Coupland (1958) are the extreme results of severe changes in the availability of moisture in a region. The more subtle physiological effects just described can be expected to result in changes in the nutritional quality of grassland forage from year to year under normal annual climatic variation.

4. SOILS AND GRASSLAND PRODUCTIVITY

The most important aspect of soil for this discussion is its nutritional value for the plants growing in it. Soils provide 14 of the 17 major nutrients required for plant growth (Thompson and Troeh 1973:13), but this discussion emphasizes nitrogen and phosphorus, the two most important of these nutrients. The quantity of nitrogen and phosphorus available to a plant is related primarily to the organic content of the soil, a major aspect of Plains soils that is known to vary systematically from region to region. Almost all of the nitrogen and about half of the phosphorus that is available to plants is derived from the organic content of the soil in which those plants are growing. The amount of these nutrients that a plant actually absorbs depends on the amount available and on the moisture in the soil: nutrient absorption generally increases with increasing nutrient availability and with increasing soil moisture.

Once a plant has absorbed a critical amount of a needed nutrient, increases in that amount result in smaller and smaller increases in plant productivity, until a point is reached where additional nutrients result in no additional growth. However, plants will generally absorb a greater amount of a nutrient if it is available and may substantially increase the concentrations of nutrients in their tissues without significantly increasing their forage yield (Macy 1936). Grasses in various regions may thus vary substantially in their nutritional content for grazing animals without varying in their vigor or overall productivity. Larger amounts of phosphorus in the soil will generally result in higher phosphorus concentrations in the plants growing in those soils. Numerous studies have also shown that increased nitrogen availability raises protein content (summarized by Lyttleton 1973:96–97; also see Pettit and Deering 1974). Both phosphorus and protein levels, then, should be higher in grasses grown in more organic than in less organic soils.

5. GRAZING AND GRASSLAND PRODUCTIVITY

Herbivores of all kinds affect grasses by continuously removing the accessible parts of the plants for food. Under extreme conditions, defoliation due to grazing pressure can severely damage a grassland, tremendously reducing total forage production, particularly when overgrazing is combined with drought (Buwai and Trlicka 1977; Tomanek 1959). However, the effects of less severe defoliation are exactly opposite to this.

All studies indicate that moderate levels of defoliation increase overall forage yield, as grasses produce new growth to compensate for that which is removed (Eck *et al.* 1975; Weinmann 1955). By repeatedly inducing new growth, grazing thus increases the proportion of the year during which grasses are in their less mature, more nutritious stages of growth, increasing the period of time when forage conditions are best. In the short term, then, grazing both increases total forage production in a region and maintains the nutritional status of the grasses at relatively higher levels than a natural grassland would attain without grazing.

McNaughton (1976, 1979, 1984) has extensively discussed the longer-term effects of grazing on grasses. McNaughton's results indicate that continuous grazing profoundly affects the productivity, composition, structure, and nutritional value of range grasses. First, they support the pattern just noted, showing a clear increase in forage production with increasing grazing pressure up to an optimum point, after which production falls off rapidly (McNaughton 1976, 1979). Grazing also leads grasses to reproduce by tillering, sending out shoots horizontally and thereby creating a short, dense canopy of plants referred to as a "grazing lawn" (McNaughton 1984:864). The species of plants selected for in such a canopy also differ from those that survive better under ungrazed conditions: taller species are particularly absent from grazing lawns (McNaughton 1979:697), a pattern also noted by Tomanek (1959) in Kansas.

Developing a grazing lawn substantially increases the concentration of nutrients in the available grasses (McNaughton 1984). When grasses are protected from grazing and allowed to grow to their maximum possible heights, they reach a point where they contain concentrations of nutrients that are too low for large grazing animals to maintain their body weights. In East Africa, this point is attained when grasses are about 40 centimeters tall. Maximum nutrient concentration, though, occurs at a higher level of grazing pressure than does maximum forage production (McNaughton 1976:696). The intensity of wildebeest grazing in the Serengeti, in fact, is above the level that is optimum for forage production, suggesting that grazing ungulates feed in ways that maximize nutrient rather than forage production.

Weinmann (1955:588–589) presents data illustrating this last point. Under several different rates of experimental clipping of South African range grasses, Weinmann found that a rate of two clips per season maximized forage production (measured in pounds of forage per acre). In contrast, protein content was highest at a rate of four clips per season, and, as a result, the total production of protein (and other nutrients) per acre was highest at this higher rate of clipping despite the lower overall amount of grass produced.

6. SUMMARY

The preceding discussion can be briefly summarized. Grasses are most nutritious in the early parts of their growing season and are reduced in quality as they mature. Seasonal patterns of growth differ from species to species, with grasses falling roughly into cool-season species, which make most of their growth in the spring and fall, and warm-season species, which make most of their growth in the summer. The process of maturation, however, is not necessarily continuous but rather occurs in direct response to the amount and timing of rainfall during the growing season. The relationship between water availability and grass growth is closest for warm-season species and weakest for cool-season species, and growth patterns are generally more erratic in warmer regions and more regular in cooler areas. These species also differ in their nutritional value, with cool-season species being more nutritious than warm-season species.

Grasslands react to reduced overall moisture availability by reducing their metabolic rates and putting less energy into growth above the ground, resulting in less carbohydrates in their leaves and lower percentages of protein, phosphorus, and other nutrients when regional moisture levels fall below the level required for transpiration. Under prolonged or extreme moisture deficits, this metabolic shift results in dramatic reductions in forage yield due to decreased plant vigor and, finally, death. Increased temperature also adversely affects forage yield by increasing the percentage of the plant composed of undigestible crude fiber and lowering its overall nutritional value. Increased precipitation produces the reverse of these patterns. In addition, the presence of greater amounts of important nutrients in the soil increases the concentration within the plant of at least some of the nutrients that are important to grazing animals.

Finally, moderate levels of grazing increase both total forage production and the nutritional value of the forage produced by maintaining plants in younger stages of growth and by modifying the composition and structure of the grassland community. This effect both increases the overall amount of nutrients available to grazing animals in a region and the proportion of the year during which maximum nutrient yields are available.

Chapter 4

Ungulate Ecology

Plains anthropologists often discuss the topic of "bison ecology" because of the obvious importance of the organization and behavior of the great herds to the human beings exploiting them, but the meaning of "ecology" in anthropological studies of bison is considerably more restricted than it is elsewhere. Although Pianka (1974:3) defines ecology as "*the study of the relations between organisms and the totality of the physical and biological factors affecting them or influenced by them*" (emphasis in original), and biological ecologists emphasize detailed studies of fertility, mortality, population growth and structure, and resource availability (Andrewarth 1961; Moss *et al.* 1982), anthropologists use the term *bison ecology* to refer almost exclusively to seasonal differences in herd sizes and movement patterns. In addition, Reher and Frison (1980) stress another important aspect of bison ecology, the density of animals in an area.

These components of bison adaptations are major determinants of the distribution of animals across the landscape in different seasons and from year to year. Unfortunately, they are difficult to study in the modern world because existing herds of bison either are confined within unnaturally small areas or do not actually live on the Great Plains.

There are at least two ways to reconstruct past bison adaptations. The first is to examine historical documents recording the locations and characteristics of bison herds on the Plains in the past in conjunction with a knowledge of modern bison ecology (e.g., Arthur 1975; Hanson 1984; Morgan 1980; Roe 1970). Although such documents are important sources of information, there are several problems in using them (these problems are discussed in more detail in Bamforth 1987).

First, the data they contain tend to be extremely impressionistic and unsystematically collected, limiting the reliability of inferences they can support.

Second, most known historical observations are difficult or impossible to place in an explicit ecological context. As is discussed later, a number of environmental factors are critical determinants of ungulate densities and movements, and information about these factors that can be related to specific records of bison herds on the Plains is often lacking (but see Moodie and Ray, 1967, for some exceptions). Last, the Historic Period, particularly the last two thirds of the nineteenth century, was a time of tremendous change in the pattern of human exploitation of the Great Plains. Documents from much of this period record observations of the bison during a time when their adaptations were being progressively disrupted by the westward expansion of white settlement and the tremendous increase in the intensity of human predation. The data they present must therefore be used carefully, and records from the mid- to late nineteenth century are likely to have limited relevance to prehistoric periods.

A second approach to reconstructing bison ecology on the Plains draws on modern ecological studies that provide general insights into the basic structure of ungulate adaptations. Many such studies rest on detailed, long-term research on many species of animals, particularly those in the Serengeti region of East Africa. This is the primary approach taken here, and this chapter discusses the data this research provides on the three aspects of bison ecology noted before: population density, migration patterns, and seasonal patterns of aggregation and dispersion.

To address these three topics, this chapter summarizes research on several more specific topics, including the general pattern of ungulate digestion and food selection, the determinants of regional population levels, the size of the area within which a herd moves within a season and over the course of an annual round, the degree of annual and seasonal herd mobility, the size of the herds present in a region, and the specific migrations they make. Taken together. these factors largely define the structure of the faunal resources within a region to which a hunting society must adapt. As the following discussion shows, patterns of forage production in a region are a primary determinant of the nature of ungulate adaptations, and this chapter focuses particularly on their effects.

1. FORAGE CONDITIONS, UNGULATE DIGESTION, AND FOOD SELECTION

To understand the effects of forage conditions on ungulate migration and aggregation patterns, it is necessary to summarize the basic means by which these animals digest the forage they consume and the effects of these digestive systems on the kinds of forage ungulates require to survive. The grasses eaten by grazing animals contain an abundance of carbohydrates, mainly within the undigestible cellulose of the cell walls, and much lower concentrations of protein, mainly in the easily digestible cytoplasm (cell contents). Ungulates do not produce the enzymes needed to break down the cellulose protecting the cytoplasm and thereby release the nutrients they require for survival. This task is accomplished by microbes that inhabit the digestive tracts of grazing animals, without which grazers would starve to death (Bell 1971; Van Soest 1982).

Bell (1971:87–88) recognizes two basic digestive strategies that have evolved to cope with the carbohydrate/protein imbalance in grasses: the ruminant system, which he characterizes as "selective," and the nonruminant system, which he characterizes as "tolerant." In a ruminant system, such as that of the bison, food is repeatedly chewed, digested, regurgitated, and rechewed, resulting in extremely efficient use of available protein. This process, though, takes a relatively long time, and this time increases in direct proportion to the amount of undigestible or minimally digestible fiber in the diet. The time needed for digestion thus sets an absolute limit on the rate at which a ruminant can process food and therefore on the total amount of food it can consume: if food is of low nutritional quality, a ruminant cannot just eat more of it to obtain the nutrients it needs to survive. Rather, a ruminant must consume food with a minimum protein content to provide for its metabolic needs; hence Bell's characterization of ruminants as "selective."

There is also a lower limit on the amount of food that a ruminant must consume. The intestinal microbes on which ruminant digestion depends need to maintain an essentially constant level of fermentation in the gut, a level that must be fueled by a relatively constant minimal amount of undigested food. When the amount of food in the system drops below this point, digestion is disrupted, and the microbes die (Church 1971:773–774). There is thus a limit on the amount of time which a ruminant can wait before feeding, which in turn limits the distance it can travel between feeding areas (Pennycuick 1979).

In contrast, nonruminants do not process food as efficiently as ruminants. Rather, they pass it through the digestive system quickly, extracting as many nutrients as possible in a relatively short period of time. Up to a point, there is thus no inherent limit on the amount of food nonruminants can consume, and they can therefore simply increase their rate of food intake to compensate for low concentrations of nutrients in the available forage. This allows nonruminants to subsist on food with nutrient concentrations that are too low to support a ruminant; hence Bell's characterization of them as "tolerant."

The quality of food that an ungulate requires is further modified by its body size. Energy requirements per unit of body mass are generally higher for smaller animals than for larger animals (Brody 1945). This means that smaller animals tend to require higher quality forage than do larger animals to maintain their body weight. However, smaller animals have lower total forage requirements because they have less mass to provide nutrients for. They thus need to spend less time eating, allowing them to spend more time searching for higher quality food. As long as forage of sufficiently high quality can be found, a smaller ruminant can survive a reduction in the total amount of food produced better than a larger ruminant (Bell 1971:91; Hobbs *et al.* 1983).

However, Bell (1971) notes that, despite these differences, all ungulates survive best on a diet of abundant forage with high nutrient concentrations and low amounts of fiber; they simply differ in their degrees of tolerance for various kinds of departures from this ideal. In response to the benefits of this optimum diet, ungulates have evolved the ability to select plants and plant parts with higher nutritional content than that of the available vegetation as a whole. They do this particularly by selecting for the plant parts that are most nutritious and by selecting plants in earlier stages of growth (see Hart *et al.* 1983; Hobbs *et al.* 1983; Jarman and Sinclair 1979; Peden 1972; Sinclair 1977).

Constraints on this selective ability result from interspecific competition, which has led various species of ungulates living in the same area to develop largely (but not completely) complementary feeding strategies (Bell 1971; Sinclair 1979; Sinclair and Norton-Griffiths 1982; Talbot 1962). These strategies rest on specializations on different species of plants and on different plant parts (such as leaves or stems) and are reflected in the different structures of the mouths and jaws of the various species. More selective feeders, which specialize on more delicate plant parts such as very young leaves, have smaller mouths, whereas less selective feeders, which consume greater amounts of more fibrous plant parts, have larger mouths (Bell 1971).

2. UNGULATE POPULATION DENSITIES

For many years, ecologists noted that herbivores of all kinds consume only a portion of the total amount of forage produced in a region and inferred from this that herbivore population sizes were not directly regulated by the amount of food available to them. More recently, Sinclair (1975, 1977) has shown that, under most conditions, the total amount of forage produced in a region is a misleading measure of potential food resources because physiological changes in plants at the end of the growing season (see Chapter 3) make their available parts unpalatable and of little or no nutritional value. Sinclair's analysis indicates that under most conditions, this seasonal decrease in forage quality provides a mechanism that directly regulates herbivore population size.

Studying the African buffalo (*Syncerus caffer*) on the Serengeti Plains in East Africa, Sinclair (1975, 1977) demonstrated that population densities and many aspects of social behavior were regulated by the scarcity of palatable and nutritious forage during the dry season. Undernutrition during periods of low food availability made individual buffalo more vulnerable to predation, reduced their resistance to disease, and resulted in the birth of calves at the beginning of the following wet season that were too weak to survive (cf. Hall 1973). As the buffalo population increases relative to the available dry-season resource base, malnutrition becomes more severe and the forces controlling population size become stronger; as it decreases, these forces become weaker.

Relying on these data, Sinclair (1975) suggests that the population size of all herbivores existing under conditions where available forage is seasonally of poor quality are similarly regulated by available food. In support of this hypothesis, Houston (1982:64–75) has shown that a similar process of population regulation occurs among American elk (*Cervus elaphus*) in Yellowstone Park (also see Barrett 1982).

As Chapter 3 discusses, the amount of forage produced in a given region is largely determined by the amount of precipitation in that region. Because the amount of forage available determines how well herbivores can get through the dry season (in the tropics) or the winter (in temperate regions), there is a direct relationship between rainfall and the number of grazing animals in a region: animal densities are higher when precipitation is higher and more forage is therefore available. Several studies (Coe *et al.* 1976; Sinclair 1975, 1977:39–42) have demonstrated this relationship in East Africa, and Teer *et al.* (1965:21–23) note that the density of white-tailed deer (*Odocoileus virginianus texanus*) in the Llano Basin of Central Texas varies directly with rainfall for the same reason.

The minimal overlap in the diets of different species of ungulates noted in the preceding section also has an effect on population densities. Although this overlap is not great, several studies (Eltringham 1974; Sinclair 1977; Sinclair and Norton-Griffiths 1982) indicate that it is sufficient to create a degree of competition for forage between ungulate species existing in a single region. By definition, this competition reduces the total number of individuals of any single species in such a region. The aggregate population of all of the competing species, though, is greater than the population any single species could attain in isolation. This is because the diverse food preferences of different species lead mixed populations of animals to consume a greater proportion of the total available plant food in a region than any single species could consume by itself (Talbot and Talbot 1963a).

3. FAMILIAR AREAS AND HOME RANGES

It is obvious that no individual member of a species exploits the entire range within which that species exists. Rather, individual animals live within a relatively circumscribed region, although the specific size and location of this region may change over time. This areal restriction results in part from physical limits on mobility, but it also reflects the adaptative advantages conferred by the undoubted ability of most animals to learn where within a region they can go to find food, water, and shelter (Arnold and Dudzinski 1978; Baker 1978). Such knowledge, for example, probably allows African buffalo to survive the dry season by concentrating their search for food in habitats where experience has taught them food can be found year-round (Sinclair 1977).

Baker (1978:387, 384) defines several concepts that describe the area used by an animal. The largest of these is the *lifetime range,* "the total area visited by an animal during its lifetime." On a smaller scale, the *familiar area* is "that portion of the lifetime range from any point in which an animal is capable of finding its way to any other point." This area may change over the course of an animal's life as it explores new areas and forgets the characteristics of other regions. Last, the *home range* is "the collection of habitats visited during a stated period"; this period may be a day, a week, a season, or longer, depending on the analysis being conducted. Baker's terminology differs slightly from that used by some other researchers: his "familiar area" corresponds to the "home range" defined by Delaney and Happold (1979:198–211) and Jewell (1966).

The distinction between a familiar area and a home range in Baker's terms rests on the fact that an animal actually uses only a portion of the area known to it during any given period of time. For example, the locations of areas containing no food resources may be known to an animal, but these areas are not likely to be occupied. The familiar area is thus the region containing the habitats that an animal regularly exploits, and it remains "familiar" because the animal periodically revisits these habitats in the course of daily or seasonal movements. For this discussion, the familiar area is of particular importance, because it defines the region within which an animal moves and thereby defines the spatial limits on that animal's migrations.

In essence, ungulates move around as much as they have to in order to obtain adequate food, water, and shelter. The familiar area is a region within which these needs can be fulfilled. Given a particular species of animal, the size of this area depends on the density and distribution of the resources in it (Delaney and Happold 1979:199; Jewell 1966). The size of the familiar area exploited by a particular member of a given species thus appears to be set by the locations and densities of its preferred foods during different seasons of the year. When these locations are scattered over large areas, the familiar area is large; when they are concentrated in smaller areas, the familiar area is small. The contrasts between the huge areas exploited by the migratory wildebeest herds and the much smaller areas exploited by the African buffalo, for example, reflect this pattern (see Maddock 1979; Sinclair 1977:133–136, 151–152), as does the contrast in the vicuna (*Vicugna vicugna*) between the small territories exploited by territorial males and the larger territories exploited by troop males (Koford 1957).

Baker (1978:399) also outlines the general selective pressures that should lead to nonoverlapping areas and socially closed herds, suggesting that there is a continuum in resource distributions affecting these pressures. At one end of this continuum, resources are universally so abundant that an animal can move anywhere it wants with no loss of fitness; at the other end, resources sufficient for an animal's year-round needs are so spatially concentrated that it is most adaptive to develop a system of territoriality, where other members of one's species are actively excluded from a bounded area. Relatively well-defined, nonoverlapping but undefended familiar areas are most likely to develop in the intermediate portions of this continuum.

In these intermediate situations, resources tend to be irregularly spread over the landscape, and their locations often change as environmental conditions change from day to day and from season to season and as these resources are consumed by the animals who live on them. In such a situation, animals that can assess accurately the condition of potential feeding areas other than the one that they are actually in have a distinct advantage over animals that cannot make such an assessment (Baker 1978:91–92, 156). However, the ability of any animal to make this assessment is reduced if more than one herd is exploiting the same area because the members of one herd may deplete the resources of one area without the knowledge of the members of the other herd. Many animals, including ungulates (Baker 1978:172), have a variety of mechanisms for obtaining the information needed for habitat assessment, one of the most important of which is simply periodically passing through and examining the various areas it exploits. The need for such explorations is much less when these areas are not used by animals of different social groups, and there is thus selective pressure for the development of nonoverlapping areas.

Variation in the size of seasonal home ranges follows the same pattern as variation in the size of familiar areas: dispersed seasonal distributions of food and water lead to large home ranges, and concentrated distributions of food and water lead to more restricted home ranges. Various ungulates illustrate this point. Wildebeest feed preferentially on very short green grass in its earliest stages of growth. During the wet season, such grass is abundant on the intermediate- and short-grass plains of the Serengeti, and the wildebeest spend this season in these areas. When this grass dries up, the herds first move northeast into areas with more permanent water and hence better forage and then north into the Mara River Valley. When the rains begin again, the herds move back onto the plains (Maddock 1979; Pennycuick 1975).

In contrast, the buffalo prefer to eat longer green grass; their large mouths prevent them from feeding on very small plant parts. During the dry season, the buffalo herds are confined to riverine areas where at least some palatable forage can be found. During the wet season, they expand their foraging areas to include the adjacent upland grasslands (Sinclair 1977:53–62). Impala also show clear seasonal changes in home range sizes, with social groups covering larger areas in the dry season when food and water are scarce and smaller areas in the wet season when food and water are abundant and widely distributed (Murray 1982).

The areas within which ungulate species move during a given season thus reflect quite closely the distribution of the kinds of food they require, and by understanding the seasonal distribution of these foods, the animals' seasonal home ranges can be predicted.

4. MIGRATION AND AGGREGATION PATTERNS

As the preceding discussions suggest, much of the structure in ungulate migration and aggregation patterns can be understood by examining the density and distribution of forage and water in the region they inhabit in conjunction with the social and reproductive needs of the animals being studied. "Migration" in this discussion is taken in the general sense proposed by Baker (1978:21) as "the act of moving from one spatial unit to another," with the specific nature of a spatial unit defined by the problem a given researcher is investigating. For present purposes, these units include the various feeding and watering areas used by an ungulate in the course of a single day as well as the general geographic units between which an ungulate may move from season to season.

In general, ungulates seem to be as mobile as they have to be in order to obtain sufficient supplies of food and water. Variation in the distance over which territorial and nonterritorial vicuna have to forage is particularly evidence of this: when nonbreeding males succeed in displacing a territorial male, they then remain within their new territory rather than traveling extensively as they did as members of a nonbreeding troop (Koford 1957). The contrast between the essentially year-round concentration of resources available to the vicuna and the widely separated seasonal resource patches available to the wildebeest also illustrates this principle.

The environment exploited by the vicuna is extremely unusual, and it is much more common for ungulates to be faced with dramatic seasonal and spatial variations in forage availability. As was discussed in an earlier section, ungulates have evolved the ability to select a diet with a significantly higher nutritional content than that of the available forage as a whole. In conjunction with this, they also have evolved the ability to seek out areas in which forage conditions are best. As is discussed in Chapter 3, a grassland at any one time is a mosaic of patches of higher and lower quality forage for a grazing animal. More nutritious food is available during the growing season in areas where rain has recently fallen and during most of the year near permanent water sources.

Some degree of mobility is an inevitable part of all ungulate adaptations simply because grazing animals deplete the food in a region by feeding on it and because water does not occur in all potential feeding areas. (Although some ungulates obtain sufficient water for their needs from the plants they eat and do not require access to other water sources [cf. Western 1975], the species of ungulates that are important here [particularly the bison] do require such access. The remainder of this discussion is directed toward this latter class of animals.)

Movements from feeding area to feeding area and back and forth between feeding areas and water sources thus comprise the majority of ungulate migrations. These movements are governed by several relatively simple principles. Ungulate migration patterns can be discussed conveniently as two topics: first, the determinants of a species' overall degree of mobility and of herd aggregation and dispersion in a given region, and second, the patterns of specific migrations.

Because many migrations occur when the forage in a local area is depleted, a herd of a given size generally moves less often when forage is abundant than when it is not, simply because greater amounts of forage take longer to eat than smaller amounts. Under poor forage conditions, animals move more often, farther, and faster than under good forage conditions (cf. Jarman and Jarman 1979; Laws *et al.* 1975:147). Similarly, when forage is widely available over large areas, the distance a herd will have to move between feeding areas will be less than when it occurs in widely separated patches. The distribution of water in a region has similar effects: if it is found in only a few locations, the forage near these locations is grazed much more heavily than elsewhere, and animals must commute between watering and feeding areas. Such commuting is reduced when water is widely available (Arnold and Dudzinski 1978:86–88; Ayeni 1975; Hall 1973; Mloszewski 1983:57; Murray 1982; Pennycuick 1979; Weaver and Tomanek 1951). In extreme cases, the distribution of water may be so restricted as to limit the range of a herd to feeding areas in the immediate vicinity of a single permanent source (Western 1975).

The characteristics of the distribution of food and water affect the size of a herd as well as its mobility. Explanations for the existence of a herding tendency among animals emphasize the protection herds afford against predation (Hamilton 1971) and the need for young animals to learn about the characteristics of their food supply from older animals (Murton 1971). Recently, McNaughton (1976, 1979, 1984) has shown that grazing in large herds also improves the nutritional quality of the grasses in the grazed areas by constantly inducing new growth and by maintaining a "grazing lawn" of shorter grasses that have higher concentrations of nutrients than taller grasses.

However, these explanations account only for the existence of a herding tendency among ungulates and do not explain the systematic variations in herd size seen from season to season within an area and from area to area within a single species. Jarman and Jarman (1979:204–205) point out that individual animals must treat the risk of predation as constant and therefore should always stay in the largest herd possible, and an analogous point might be made about the beneficial effects of herd size on learning and forage conditions.

Seasonal and regional variations in group size are linked to the same factors as the degree of animal mobility discussed before. Ungulates generally aggregate into larger herds when food is locally abundant than when it is not, a relationship that holds for seasonal changes in food abundance within an area and for differences in abundance between areas during a single season. Under the typical pattern of seasonal forage production, herds tend to be largest during the growing season and smallest during the dry season or winter. Water availability also influences the number of animals in a given area, with animals tending to form relatively large aggregations near water sources during all seasons when these sources are few in number, as they usually are, and dispersing when they are more widely distributed (Arnold and Dudzinski 1978:73–81; Ayeni 1975; Baker 1978:547–551; Hall 1973; Jarman and Jarman 1979; Mloszewski 1983:51–59; Sinclair 1977:141–142; Western 1975).

These basic relationships can be modified by winter conditions in temperate regions (Baker 1978:551). During the winter, snow can satisfy an animal's need for water, removing the concentrating effects of a restricted distribution of water sources and enhancing the dispersive effects of seasonally poor forage conditions. However, snow that is too deep for an animal to dig through can force ungulates to seek out the often restricted wind-swept or sheltered areas where forage is accessible (e.g., Barrett 1982:994), resulting in relatively large aggregations of animals in these areas. Aggregations can also provide protection from extreme winter cold and wind.

Gregarious ungulates also aggregate for social reasons, particularly for the rut. However, forage conditions limit the size of all aggregations, including those formed for reproductive purposes. The fact that aggregations associated with the rut often occur during the season of greatest forage production indicates this relationship clearly. Among African buffalo, for example, group size begins to increase before the rut begins, synchronized with increasing rainfall and hence increasing forage production (Sinclair 1977:142).

Overall, herd size and mobility are strongly linked to the regional distributions of forage and water. Gregarious animals seem to live in the largest groups possible, with the upper limit on herd size set by these distributions: that is, herds tend to be larger and more sedentary when food is widely abundant than when it is not. A concentrated distribution of water also leads to aggregation, whereas a wider distribution of water tends to lead to dispersion. Relatively light snowfall can enhance the dispersion expected under poor forage conditions, but excessive snowfall or winter cold and wind may lead to aggregation despite a relative lack of food.

In addition to showing that it is possible to predict general levels of ungulate mobility and aggregation, many studies have shown that it is often possible to predict specific migrations. This is true not only for large-scale seasonal movements between widely separated habitats, such as those carried out by the Serengeti wildebeest, but also for smaller movements between feeding areas within a season. A variety of evidence indicates that many animals, including ungulates, perform "calculated migrations" (Baker 1978:25) based on assessments they make of the suitability of a given habitat relative to the habitat they are in. Thus, when local forage is depleted, animals often move to specific locations where they expect feeding conditions to be better.

Baker (1978:91) suggests three major sources of information on which an animal can base such an assessment: "direct perception, previous acquaintance, and social communication." In the first of these, animals may either directly perceive the condition of a habitat (by scent or sight, for example), or they may assess a perceptible indirect measure indicating that condition. The second potential source of information refers to the knowledge of the characteristics of a familiar area or habitat gained over the life of an animal and to the short-term information on habitat suitability gathered during day-to-day movements within a home range (cf. Pennycuick 1979:182). The third refers to patterns of behavior by some members of a herd that let the other members of the herd know where food is. Baker (1978:165–172) has shown that all of these strategies seem to be used by various species of ungulates.

Perhaps the most spectacular, calculated ungulate migrations on record are made by the wildebeest, who watch for distant rainstorms during the dry season and move to them when they see them (Talbot and Talbot 1963b). Buffalo also sometimes migrate to rain at the very beginning of the wet season but more often move out of their dry-season riverine habitats onto the adjacent plains 2 months after rain has fallen and the grasses have had time to grow (Sinclair 1977:66–67, 98). Under normal conditions, impala defend clearly bounded territories against conspecifics but will abandon these territories in the dry season to move to adjacent areas that have received rain (Jarman and Jarman 1979:210). During the dry season, all species of water-dependent animals in the Amboseli region of Kenya leave the few permanently wet areas where they concentrate to move into adjacent areas within hours of even a light rain (Western 1975). Animals in herds within a bounded home range also know where they can expect to find forage at different times of the year and also where they have grazed recently, allowing them to predict areas that are likely to be productive. Similar patterns to these can be expected for seasonal changes in the locations of water.

Finally, competition between species generally tends to disperse animals over the landscape. It does this in two ways. First, the complementarity between the feeding strategies of different species of ungulates discussed earlier tends to create a "grazing succession" (Bell 1971; Pennycuick 1979). In this succession, a species with one food preference grazes an area and, by this grazing, makes the area more suitable for a species with another preference. This second species then moves into the area after the first has left. For example, a species that eats mainly leaves may graze in a region and deplete this preferred food. Removing the leaves exposes the stems and improves grazing conditions for a species that prefers them. An area that was no longer suitable for the first species is thus suitable for the second, and at least one species of animal is present in the area for a longer period of time. Second, different species of large ungulates inhabiting the same area tend to avoid each other at waterholes (Ayeni 1975), which should inhibit the degree of concentration of different species of animals at such locations.

5. SUMMARY

Overall, it is clear that many important aspects of ungulate ecology can be understood by examining regional distributions of forage and water. These include the density of animals in an area, the size and boundaries of the areas within which they migrate, the size of the herds they move in, their overall degree of mobility, and their specific migration patterns at a given time.

Ungulate population densities are determined by regional forage production that in turn is determined by rainfall; densities are higher in areas with more available forage and lower in areas with less. In addition, under most conditions ungulates will move over the course of a year within a reasonably well-bounded familiar area whose size depends on the seasonal distribution and density of their preferred forage. These areas, and the more restricted portions of them exploited during any single season, are larger when forage is scarce and smaller when it is abundant. Herd size similarly reflects local forage abundance, being larger when more food is available. A restricted distribution of water in a region also tends to create relatively large aggregations of animals, whereas a dispersed distribution of water tends to create dispersed distributions of animals. However, both this effect and that of forage abundance can be modified by excessive snowfall and severe winter conditions, which may lead to aggregation when other factors would predict dispersion. Seasonal and annual mobility also increase as forage becomes scarcer. Finally, specific movements by a given herd are often calculated to bring them to areas where they expect tc find food and water, with calculations often made by knowledge of the distribution of rainfall and previously grazed areas within a home range.

Competition between species coexisting in an area also affects these patterns. This competition reduces the population of any single species but increases the aggregate population of the groups of competing species. It also disperses animals of different species over the landscape by leading them to use particular feeding areas and watering places at different times.

Taken together, these patterns suggest strongly that a hunter can anticipate animal distributions and densities with a fair degree of accuracy on the basis of factors (such as rainfall distributions) that are readily perceptible and easily learned. This does not mean that any ungulate is a perfect information-processing machine whose every action is optimally designed and hence totally predictable. However, it does mean that there is a rational basis on which hunters can rely in deciding when and where to look for game.

Chapter 5

Patterns of Forage Production on the Great Plains

Chapter 2 proposed that more heterogeneous hunting societies on the Plains should be found in regions in which ungulates occur in larger, more widely separated herds that are less mobile and whose movements are relatively regular within a season and repetitive from year to year. To test this proposition for the recent bison-hunting societies on the Plains, it is necessary to consider the nature and degree of variation in the determinants of bison adaptations there. As Chapter 4 shows, patterns of forage production are the most important of these determinants, and forage production is largely controlled in turn by climatic factors. This chapter therefore considers the differences in modern climate across the Plains and demonstrates the effect of these differences on forage production.

1. CLIMATIC VARIABILITY ON THE GREAT PLAINS

This summary of the nature of climatic differences across the Great Plains draws largely from Borchert (1950), Court (1974), Hare and Hay (1974), and Kincer (1923). Its emphases follow from the discussion in Chapter 3 that showed that forage production is determined by seasonal and spatial patterns of precipitation and temperature and from the discussion in Chapter 4 of the effects of winter conditions on ungulate adaptations.

Mean annual precipitation decreases to the west across the Plains from a high of roughly 100 centimeters (40 inches) to a low of 30 centimeters (12 inches) adjacent to the Rocky Mountains. The major exception to the pattern is in the area in and around the Black Hills of eastern Wyoming and western South

Dakota, where annual precipitation is substantially higher than in the surrounding region. The degree of year-to-year variation in total precipitation as measured by its coefficient of variation (CV) increases from northeast to southwest, from approximately 20% on the far Northern Plains in Canada (Hare and Hay 1974, Figure 22) to over 40% in southeastern New Mexico (Hershfeld 1962). In addition, the CV increases much faster across the Southern than the Central and Northern Plains (Hershfeld 1962, Figure 2).

Most rain on the Plains falls during the spring and summer, but the pattern of seasonal variability in rainfall is somewhat different in different areas. Thornthwaite (1941:179–180) has examined the frequencies of seasonal droughts on the Northern, Central, and Southern Plains. His data indicate that the South Plains are more drought-prone overall than elsewhere and that this is particularly true for the winter. In addition, South Plains winters are substantially more likely to have longer droughts than other seasons: in more northern areas, such seasonal differences are less pronounced. Precipitation on the Northern Plains also tends to be more evenly distributed throughout the year. On the average, there are only approximately 60 days per year with at least 0.25 mm of precipitation in the Texas Panhandle, but 90 to 110 such days in Montana and North Dakota, and although 60% to 70% of the total annual precipitation in the former area falls during storms that deposit at least 0.5 inches of rain, only 30% to 50% falls in such storms in the latter region.

Other data indicate that there are systematic differences in the spatial distribution of precipitation on the Northern and Southern Plains. Longley (1974) and Haragan (1976) have studied the correlation in monthly precipitation between weather stations on the Canadian prairies and Texas High Plains, respectively, particularly in relation to the distance between pairs of stations. This correlation falls off as the distance increases in both areas, but it does so faster in the south than in the north. Table 5-1 compares the distance at which the mean correlation between stations falls below 0.6 for each month in the growing season in Canada and Texas. In both regions, the distance fluctuates substantially from month to month, but the overall pattern in all months but July is for precipitation to be more similar over a larger area in the north than in the south.

On the average, the first snow falls on the Canadian plains by the beginning of November and the last falls in the middle of April; comparable dates on the Texas Plains are mid-December and mid-February. Mean annual snowfall in the north is as much as 48 inches, four times the amount that falls in the south. The greatest number of continuous days with below-freezing temperatures recorded in North Dakota between 1950 and 1960 was 50, compared with a maximum of 6 days near Amarillo. The mean length of the frost-free period on the Plains ranges from approximately 100 days in southern Saskatchewan to 210 days near Lubbock, Texas. Osborn's (1983) data on winter severity on the Plains show variable but generally hard winters on the Northern and Central High Plains with

Table 5-1. Distance (in Miles) at Which the Monthly Correlation in Rainfall between Weather Stations during the Growing Season Falls below 0.6 on the Northern and Southern Plains[a]

Month	Northern Plains	Southern Plains
April	181.8	175.4
May	127.3	107.8
June	163.6	61.2
July	45.5	130.2
August	109.1	48.3
September	190.9	125.5

[a]From Haragan 1976 and Longley 1974.

a relatively abrupt decrease in severity from the central plains of Kansas (winter severity index = 19.15) to the Southern Plains of Texas (winter severity index = 3.84).

Temperatures are always lower in the north than in the south, but the maximum differences are most pronounced in the winter. January temperatures average roughly 45 °F lower in Alberta and Saskatchewan than on the Southern Plains, whereas the comparable difference in July temperatures is only about 15 degrees. Probably because of these lower temperatures, evaporation rates in the north are substantially lower than in the south.

Karl and Koscielny (1982) examined the spatial pattern of correlation in drought severity in the United States from 1895 to 1981 and were able to define nine relatively clear regions of the country within which droughts tended to occur at the same time; correlations between these regions were uniformly low. The Plains fall mainly into two of these regions, one containing Texas, Oklahoma, and part of Kansas and Colorado, and another containing northern Colorado, Nebraska, Wyoming, North and South Dakota, and Montana. A closer examination of the pattern of correlations shows one center of relationship in Texas and Oklahoma and another in Wyoming, Montana, western Nebraska, and North and South Dakota; the intermediate area is about as strongly related to one as to the other. This pattern indicates that, during the period Karl and Koscielny examined, climatic events on the Southern Plains tended to be relatively independent of such events on the Northern Plains. Similarly, Vines (1982:7304) found that rainfall patterns in the Northern, Central, and Southern Plains tended to vary independently of one another.

The climatic distinctions between the Northern and Southern Plains appear to result from differing patterns of airflow in the two regions. Southwesterly winds on the Southern Plains carry weather out of the mountains in Mexico and New Mexico; northwesterly winds in the north carry weather out of the northern

Rockies. These two air masses meet over southeastern Colorado and Kansas. This pattern is apparent in Borchert (1950) and explicit in Bryson *et al.* (1970). These latter authors argue that the different sources of weather could have led to opposite responses on the Northern and Southern Plains to some climatic changes, with precipitation increasing in the south and decreasing in the north under certain conditions.

A major effect of these patterns is a constant increase in the difference between the amount of water available to plants and the amount of water they need to reach their full growth potential as one moves from northeast to southwest across the Plains. Measurements of this difference, the total annual water deficit in a region, have been compiled by Mather (1964) and are listed for a representative sample of weather stations on the Plains in Table 5-2. Figure 5-1 depicts the distribution of these values within the Plains, showing the pattern just noted.

To summarize, an east-to-west decrease in precipitation together with a

Table 5-2. Total Annual Water Deficit (in Millimeters) at Selected Weather Stations on the Great Plains[a]

Station	Deficit
Lethbridge, Alberta	172
Lamar, Colorado	357
Sterling, Colorado	235
Hays, Kansas	164
Topeka, Kansas	33
Harlem, Montana	268
Miles City, Montana	304
North Platte, Nebraska	210
O'Neill, Nebraska	87
Clovis, New Mexico	324
Roswell, New Mexico	510
Dickinson, North Dakota	177
Jamestown, North Dakota	74
Boise City, Oklahoma	159
Tulsa, Oklahoma	85
Regina, Saskatchewan	162
Saskatoon, Saskatchewan	166
Gettysburg, South Dakota	244
Rapid City, South Dakota	170
Abilene, Texas	378
Amarillo, Texas	268
Midland, Texas	598

[a]Data from Mather 1964.

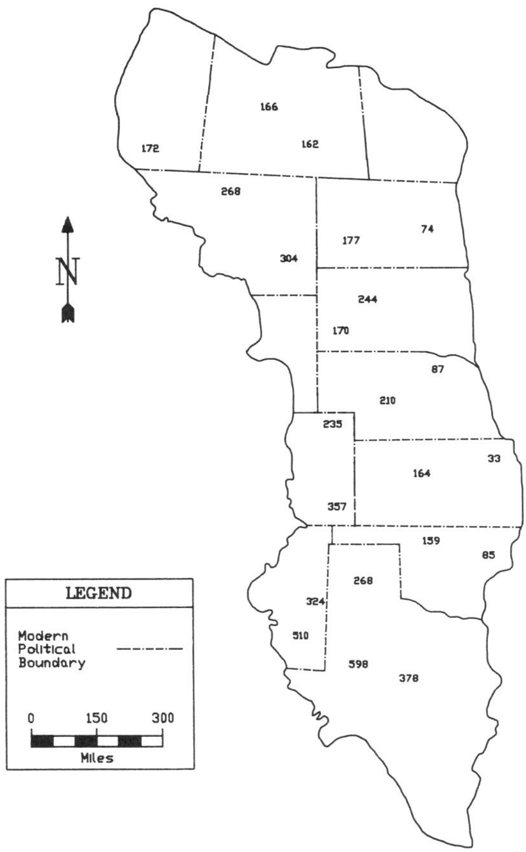

Figure 5-1. Total annual water deficit (in millimeters) at selected weather stations on the Great Plains.

north-to-south increase in temperature and thus evaporation rates create a major northeast-to-southwest decrease in effective moisture across the Great Plains except in the Black Hills, where effective moisture increases substantially. Furthermore, this southwestward decrease is accompanied by a substantial increase in year-to-year variation in precipitation. Less abundant rainfall is also more patchily distributed in space, and the decrease in precipitation across the Plains should therefore be paralleled by an increase in its patchiness at any given moment; monthly rainfall is clearly patchier in the south than in the north. The rain that falls in the south is also more episodic than in the north, particularly in the summer. Temperatures during all seasons are higher in the south than in the north. As might be expected because of this, the Northern Plains have much more severe winters with more snow, earlier first snows and later last snows, more severe extreme winter conditions, and colder winter temperatures overall than do the Southern Plains.

2. FORAGE PRODUCTION ON THE GREAT PLAINS

Chapter 3 indicates that climate, soils, and grazing intensity are the major determinants of the amount and quality of forage produced in a region. This chapter discusses only the effects of the first two of these, on the grounds that the pressure exerted by bison and other herbivores on the Great Plains before the mid-1800s was probably more or less constant as a result of a long period of mutual adaptation of grasses, herbivores, and the herbivores' predators (including human beings).

Detailed descriptions of the vegetation of the Great Plains as a whole and of specific regions within them are available in many sources (Allred and Mitchell 1955; Coupland 1961, 1979; French 1979; Sims *et al.* 1978; Weaver 1954; Weaver and Albertson 1956; Weaver and Clements 1938:520–529). The Plains grasslands are generally divided into four types based on the dominant species in them. Grasslands dominated by tall grasses are referred to as true prairie and occur from Texas into Manitoba, between the Eastern Woodlands and approximately the 100th meridian (Figure 5-2). West of the true prairie, the grasslands are dominated by mid- and short grasses. In the past, this entire area has been referred to as the mixed prairie (Weaver and Clements 1938; Weaver and Albertson 1956) despite the domination of its western and southern portions by short grasses; this domination was explained as the result of historic overgrazing. More recent research (French 1979; Sims *et al.* 1978) recognizes this domination as a natural condition and divides the High Plains into an eastern mixed prairie section and a western short-grass prairie section. These associations extend from the 100th meridian west to the Rocky Mountains and from Saskatchewan into central Texas. In the more arid areas of the southwestern United States and Mexico, Desert Plains Grasslands dominated by short grasses mixed with small shrubs and succulents occur.

The general discussion of grassland ecology in Chapter 3 indicates that systematic variation in precipitation, temperature, and soil fertility within the Plains should lead to systematic variation in forage yield and quality. To begin, it is useful to examine the overall, or "average," forage conditions given the basic differences in climate and soils in different parts of the Plains. This examination can then provide a basis for understanding the patterns of deviation from average conditions. On the average, forage should be more abundant and nutritious in those regions of the Plains that experience less severe water deficits for shorter

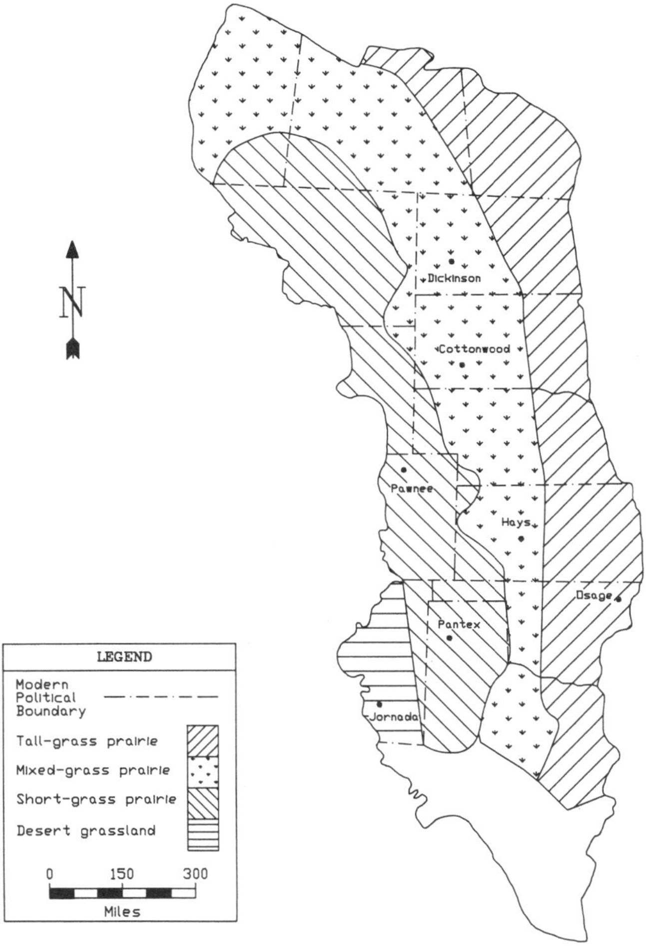

Figure 5-2. Major grassland types on the Great Plains.

portions of the year, that have lower growing-season temperatures, and that have more fertile soils.

The preceding section showed that there is a substantial decrease in the quality of growing conditions across the Plains, with the more southern and western regions having the highest seasonal temperatures, most severe annual water deficit, and most prolonged period of the year during which available water is deficient for transpiration. The clear implication of this is that, under natural conditions, the grasslands of the more southern and western Plains should have the lowest forage yield and nutrient content of any area in the region. In addition, the patchy and erratic distribution of rainfall within these more arid regions should result in more erratic growth patterns and hence in greater within-season variation in forage quality in any given local area.

Relatively detailed data provided by Aandahl (1982) strongly support this prediction about regional differences in total forage production within the Plains. As part of his exhaustive description of the various soils of the Great Plains within the boundaries of the United States, Aandahl estimates their potential native forage production in pounds of dried grass per acre. To use these data here, a 50-mile grid (an interval chosen for convenience) was laid down over Aandahl's map of soil units on the Plains, the soil types found at each grid point were noted, and the potential forage production for those soil types was recorded. Where different values for potential production were given for several soils occurring in a single unit, the value used here is the midpoint of the range of these values.

Figure 5-3 is a contour map of forage production on the Plains based on these data, produced using the SYMAP mapping program (Dudnik 1971). The total range of data values was divided into seven intervals that attempt to conform to natural breaks in the data while keeping both the number of points and the range of values at least roughly comparable from interval to interval. Two of the grid points fell into the Black Hills, which Aandahl (1982:xv) does not consider to be pedologically part of the Plains and so does not describe. The values used for these points are those for soils formed at the same latitude under similar amounts of rainfall to those in the Black Hills.

The total forage production depicted in Figure 5-3 obviously decreases dramatically from east to west and less dramatically from north to south. The major exception to this overall pattern is a notable peak in forage production in the Black Hills region of eastern Wyoming and western South Dakota. Table 5-3 shows the frequency of grid points in each level, dividing the mapped area into northeastern, northwestern, southeastern, and southwestern portions by drawing a line from east to west at approximately the Kansas–Nebraska border and from north to south at approximately the 100th meridian. This table departs substantially from random expectations ($\chi^2 = 218.5$, $df = 18$, $p > .001$) and demonstrates not only the lower productivity of the western region but also that the southern portions of both the eastern and western regions do not reach the maximum productivities of the northern portions.

The fertility of Plains soils follows a similar pattern to that shown before for precipitation because of the close relation between fertility and soil water availability. The two most important nutrients in soils are nitrogen and phosphorus, with almost all of the former and about half of the latter coming from decomposing organic matter. More arid regions have less organic matter and less effective decomposition, resulting in less of these nutrients and hence lower fertility (Thompson and Troeh 1973:126–128). Although Plains soils as a whole are very

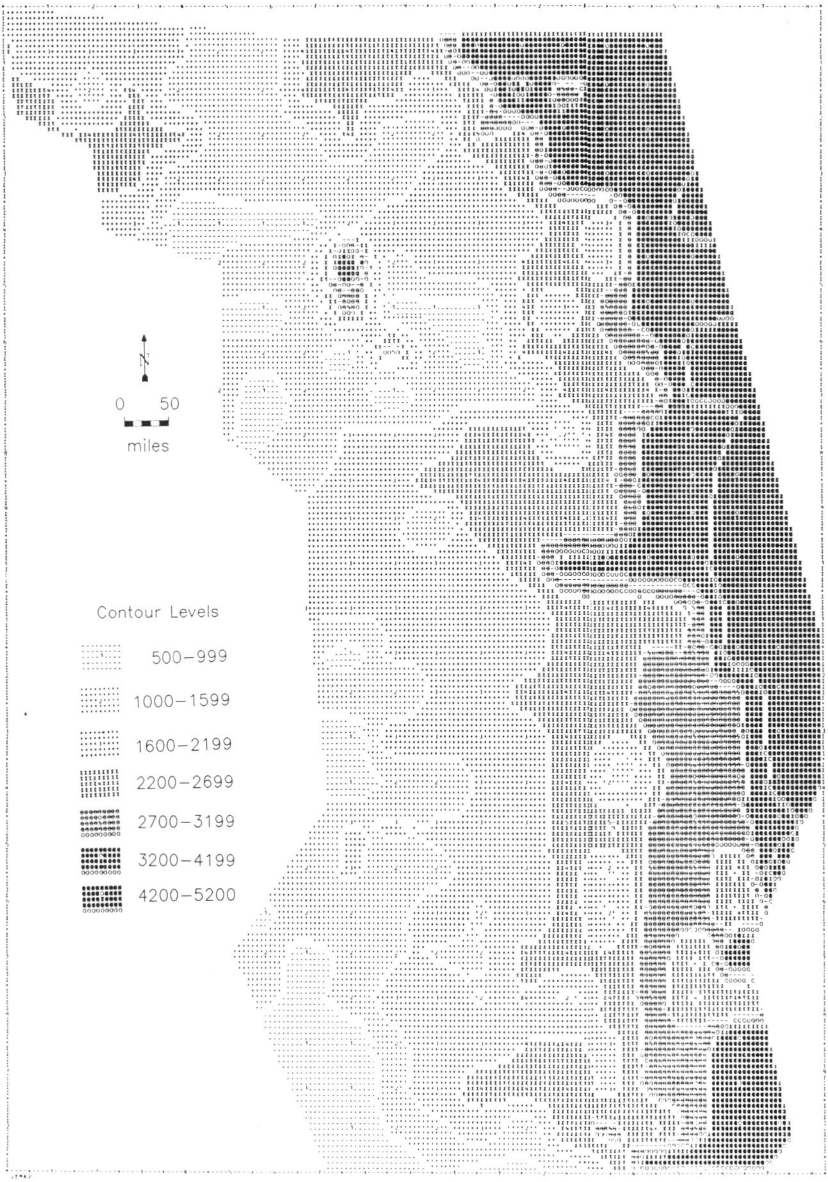

Figure 5-3. Forage production (lb dry grass/acre) on the United States Great Plains.

Table 5-3. Frequencies of Grid Points in Figure 5-3 by Level of Forage Production and Map Quadrant

Area	550–999	1000–1599	1600–2199	2200–2699	2700–3199	3200–4199	4200–5200	Total
NE	0[a]	1	3	17	6	26	9	62
	(5.2)[b]	(12.5)	(13.1)	(14.1)	(6.4)	(7.6)	(3.0)	
NW	13	33	35	29	3	3	1	117
	(9.9)	(23.6)	(24.7)	(26.6)	(12.2)	(14.4)	(5.6)	
SE	0	5	3	18	23	9	5	63
	(5.3)	(12.7)	(13.3)	(14.3)	(6.6)	(7.8)	(3.1)	
SW	13	23	24	6	0	0	0	66
	(5.6)	(13.3)	(13.9)	(15.0)	(6.9)	(8.1)	(3.2)	
Total	26	62	65	70	32	38	15	308

[a]Observed frequency.
[b]Expected frequency.

fertile when compared to other types of soils, there are substantial differences among them. Soils in the southern and western regions are drier than those elsewhere and therefore have less organic matter and hence less available nitrogen and organic phosphorus (Aandahl 1982). Overall phosphorus levels in the Plains decrease from north to south with a local peak in the vicinity of the Black Hills (Thompson and Troeh 1973:276–277). Like the climatic patterns just discussed, these patterns should result in lower concentrations of protein and phosphorus in grasses growing under natural conditions in the southern and western regions than in the northern and eastern regions.

Data on variation in the nutritional value of grasses on the Plains in response to these differences are more difficult to obtain. Many studies with information that might be relevant to this topic express nutritional value in noncomparable units and do not present the data needed to convert them to a common measure or do not specify the environmental conditions under which the plants they studied grew. This last problem is particularly important because of the great variability in weather on the Plains: comparing grasses grown under exceptionally moist conditions in a normally arid region to grasses grown under exceptionally arid conditions in a normally moist region would be misleading. All of the data discussed here should therefore be taken to indicate overall trends rather than precise quantitative differences.

The only aspect of forage nutritional value on which adequate comparative data exist is crude protein content, generally expressed as the percentage of the total plant composed of protein. This emphasis is due mainly to the importance of protein to the diet of grazing animals (e.g., Sinclair 1977; Speth 1983; Van Soest 1982). Table 5-4 presents two perspectives on the published data on

Table 5-4. Crude Protein Content of Several Native Range Grasses on the South Plains, in Wyoming, and on Average in the United States

Species	Average[a]	South Plains[b]	Wyoming
Western wheatgrass	17.0	10.6	14.3[c] 18.5[d]
Sand dropseed	12.2	11.0	—
Sideoats grama	11.6	10.2	—

[a]National Research Council 1971.
[b]Willard and Schuster 1974.
[c]Rauzi *et al.* 1969.
[d]Hart *et al.* 1983.

protein levels in native range grasses. The first is a comparison of the measured protein content of one grass species (western wheatgrass) in Texas and Wyoming, areas with comparable levels of rainfall but different overall temperatures and hence different evaporation rates. The second is a comparison for three species of grasses of their protein content in West Texas with mean protein contents computed on data from grasses grown in a variety of unspecified locations in North America (average values were taken from National Research Council 1971).

These values represent the crude protein content of the aerial parts of the grasses in the earliest stages of growth on which data were available. If the phenological stage was specified, the data for immature plants were the first choice, with data for early bloom plants used if these were not available. If the phenological stage was not specified, the value from the earliest part of the growing season that was recorded was used; in all cases, this was no later than early June. The protein content of western wheatgrass is notably lower on the Texas High Plains than in Wyoming. In addition, the protein content of all three species examined on the South Plains is lower than the average, although not as dramatically.

As is discussed before, there is a basic climatic difference between the Plains north and south of the region around southern Colorado and the Oklahoma Panhandle. South of this region, the climate is dominated by air from northern Mexico and the southern Rocky Mountains; north of it, the climate is dominated by air from the Pacific Northwest and the northern Rocky Mountains. These two regions differ particularly in the degree of variability in annual rainfall, with the Southern Plains having much greater deviations from average than the northern region. Forage production from year to year in the south, then, must follow this pattern, varying dramatically and unpredictably. Such variation must be much less in the northern region.

Within a given year, it is also important to consider the spatial distribution of rainfall and hence grass growth in a region. In general, lower amounts of rainfall tend to be more patchily distributed over a region than higher amounts of rainfall: when more rain falls in a year, it tends to be more evenly distributed. Unfortunately, there are no data available on the spatial distribution in different parts of the Plains of forage of varying quality at a single point in time, and these distributions must therefore be extrapolated from rainfall patterns. Working in the Serengeti grasslands of East Africa, McNaughton (1979:53–57) has shown that such an extrapolation is reasonable.

Because Plains grasses grow in direct response to local rainfall, the grasslands at any specific point in time can be seen as a mosaic of patches containing varying amounts and qualities of forage, with the amount and quality of forage increasing in a patch when rain falls on it. The more localized nature of growing-season rainfall on the Southern than on the Northern Plains indicates that high-quality patches in this mosaic should be smaller and more randomly dispersed in the south than in the north. Furthermore, the relatively low spatial correlation in rainfall on the Southern Plains indicates that changes in the forage conditions in one patch will not be closely related to changes in forage conditions in other patches. This is true for both the growing and the nongrowing seasons.

Growing season temperature also varies across the Plains, although not as drastically as precipitation. This variation is sufficient, though, to alter the frequencies of cool- and warm-season species in different parts of the region. As Chapter 3 discusses, warm-season grasses are more abundant in warmer climates and cool-season species are more abundant in cooler climates. Following this pattern, Teeri and Stowe (1976) have shown that the number of warm-season species in grassland communities increases from north to south across North America, with roughly twice as many of these species present in Texas and Oklahoma as in North Dakota. Similarly, Sims *et al.* (1978:257) have studied ten North American grasslands and, just as this relationship predicts, note that those in warmer areas tend to be more dominated by warm-season species than those in cooler areas.

Higher proportions of late-blooming warm-season species on the South Plains means that the bulk of forage production should occur later in the year than in more northern areas, and Sims and Singh (1978b:557, 560–562) show that this is the case. These authors' plots of forage production over the growing season (Sims and Singh 1978b, Figures 1–10) also show that substantial amounts of forage are produced during less of the growing season on more southern grasslands than more northern grasslands, implying that poor winter forage conditions tend to persist longer in the south than in the north. Table 5-5 shows estimates derived from these plots of the percentage of the growing season during which the amount of forage produced is at least 25% of the maximum amount. Because warm-season species are also less nutritious than cool-season species, their dominance on the Southern Plains also supports the likelihood noted before

Table 5-5. Percentage of the Growing Seasons during Which Forage Production Is at Least 25% of the Maximum Attained during the Year at Seven Experimental Grasslands on the Great Plains[a]

Grassland	Percentage
Dickinson, North Dakota	80
Cottonwood, South Dakota	80, 75, 75
Pawnee, Colorado	100, 100, 100
Hays, Kansas	65
Osage, Oklahoma	58, 56, 56
Pantex, Texas	5, 37, 60
Jornada, New Mexico	54, 20, 45

[a]Arranged from north to south. Multiple values refer to measurements made during different years (from Sims and Singh 1978a). The locations of these grasslands are noted in Figure 5-1.

that the grasses in this region tend to be less nutritious than those found elsewhere. Although the number of years over which observations were made is small, these data also show the greater degree of year-to-year variation in forage conditions expected on the Southern than the Northern Plains.

3. SUMMARY

Overall, then, there is a substantial decrease in the amount and probably the quality of forage produced on the Plains from east to west and from north to south, with the exception of a peak in production within the Northwestern Plains in the Black Hills. Frison (1978:191–192; also see Dodge 1965) has similarly noted the better forage conditions in this area. In addition, in more arid areas, more abundant and higher quality forage should be more patchily distributed, with the locations of more productive patches changing less predictably in response to local rainfall than in moister regions. Fluctuations in total forage production from year to year as a response to fluctuations in annual precipitation are substantial throughout the Plains but should be much greater in the south than in the north. The greater dominance of later-blooming warm-season species in southern than northern grassland communities also indicates that the bulk of forage production in the south begins later in the year and continues for a shorter period of time than in the north. This dominance also implies that the pattern of grass growth through the year will be more erratic in the south and more continuous in the north because of the closer relation of growth to local rainfall in warm-season grasses.

Chapter 6

Eighteenth- and Nineteenth-Century Climate and Bison Adaptations on the Great Plains

The preceding chapter summarized the modern pattern of climatic variation on the Plains and assessed the effects of this variation on regional patterns of forage production. Much of this variation is linked to the physiography of the area and is therefore likely to have been relatively constant over time. A north-to-south increase in temperature (although not the specific temperatures recorded in the twentieth century, of course), for example, can reasonably be assumed for all periods of human occupation. However, some evidence suggests that there may have been important differences between the climate of the eighteenth and nineteenth centuries and that of the present. This chapter discusses these differences, predicts the effects on bison adaptations of variability in the Plains grasslands, and considers modern and historic evidence to test these predictions.

1. THE EFFECTS OF THE LITTLE ICE AGE ON THE PLAINS GRASSLANDS

Climatically, the period from approximately the fourteenth century until the mid-nineteenth century (see Lamb 1977:463) is usually referred to as the Little Ice Age. This period was characterized on a global basis by lower overall annual temperatures and increased annual rainfall compared to the present (Lamb 1966:65–66, 1977:461–473), a reconstruction to which Plains anthropologists often appeal (i.e., Osborn 1983; Reher and Frison 1980:40). However, there are two good reasons to consider the evidence for these changes on the Plains in some detail. First, Lamb (1977:400–402) documents variation in local climatic conditions within the overall Little Ice Age pattern, implying that it is necessary to show that "Little Ice Age" conditions actually obtained in any specific region. Second, this general description tells us very little about the specific seasonal changes in temperature and precipitation that might have created systematic differences between the grasslands of the eighteenth and nineteenth centuries and those of the present; more detailed data are needed to understand the effects that climatic change during this period might have had on the Plains.

It is possible to identify two basic kinds of more detailed reconstructions of recent Plains climate. The first depends heavily on data on general modern atmospheric circulation patterns and global paleoenvironmental data in order to draw inferences about past circulation patterns and the local climatic conditions these patterns might have created (Bryson and Wendland 1969; Sanchez and Kutzbach 1974). The second kind is more empirical and relies on historic meteorological records from the Plains (Wahl and Lawson 1970) or other data from which climatic conditions can be inferred, such as historical diaries or other documents (Lawson 1974:33–76; Rannie 1983) and tree rings (Blasing and Fritts 1976; Duvick and Blasing 1981; Fritts *et al.* 1979, 1981; Lawson 1974:7–32; Stahle and Cleaveland 1988; Stockton and Meko 1983).

The differences between these two basic kinds of analyses are important because they have sometimes led to at least partly conflicting reconstructions. The empirical reconstructions are taken here to be more reliable than those based primarily on deductive models for two reasons. First, the factors determining global circulation patterns are extremely complex and relatively poorly understood at present, and models of their interactions appear to be particularly problematic at time scales from 10 to 1,000 years, exactly the range that is important here (Kutzbach 1976). Second, the climate of any given region is determined by a complex set of factors in addition to atmospheric circulation patterns and can be predicted only partially on the basis of these patterns alone (Bradley 1985:15–24). This implies that inconsistencies between deductive predictions and empirical data are more likely to indicate areas where the models need to be improved than problems with the data.

Tree-ring studies and historical data all indicate that Little Ice Age winter temperatures were colder on the Great Plains, with the decrease in winter temperatures apparently greatest on the northeastern Plains and progressively less to the southwest (Blasing and Fritts 1976; Fritts *et al.* 1979, 1981; Rannie 1983). Detailed analysis of tree rings (Fritts *et al.* 1979, 1981) indicates further that modern and overall Little Ice Age summer temperatures on the Plains differed little, if at all. The tree-ring data also suggest that there was much greater year-to-year variation in temperature than at present (Fritts *et al.* 1979).

Historical meteorological data collected between 1850 and 1870 suggest slightly cooler summer conditions during those two decades (Wahl and Lawson 1970), which seems to contradict the tree-ring reconstructions. However, two problems indicate that these data may be somewhat misleading. First, the period from 1850 to 1870 is often placed within the period of transition to modern climate rather than within the Little Ice Age itself (e.g., Bryson and Wendland 1969:280), making its relevance to the preceding 400 years somewhat questionable. Second, we must be extremely cautious in taking data from a period of only 20 years to represent conditions over several centuries in an area whose climate is a variable as that of the Great Plains. Evidence on long-term regional precipitation patterns (discussed later) suggests that this second problem casts considerable doubt on Wahl and Lawson's conclusions.

The usual reconstruction of Little Ice Age precipitation on the Plains sees a significant increase relative to modern normals. Reher and Frison (1980:40). for example, assert that the Plains were wetter overall at this time but present no evidence that this was actually the case. Inferring that the climate of the 1960s was similar to that of the Little Ice Age, Sanchez and Kutzbach (1974) indicate that winter precipitation on most of the Plains may have been higher than at present except in southeastern New Mexico, but they present no data from the Plains that specifically support this reconstruction. The major evidence supporting these reconstructions derives from Wahl and Lawson's (1970) meteorological data from the 1850s and 1860s, which show increases in precipitation relative to modern conditions throughout the Plains as high as 20%. The discussion here notes that there is some reason to question the relevance of this period of time to the Little Ice Age as a whole, but there are stronger grounds for treating this reconstruction with caution.

These grounds derive from studies of long-term tree-ring sequences that span not only the Little Ice Age itself but also the 20-year period for which Wahl and Lawson were able to obtain information; these sequences also continue into the twentieth century. The most enlightening of these studies for the historic meteorological data is Duvick and Blasing's (1981) tree-ring-based reconstruction of precipitation in Iowa from 1680 to 1979. Wahl and Lawson (1970, Figure 7) reconstruct an increase in precipitation relative to modern normals, defined as precipitation from 1931 to 1960, in the Iowa region on the order of 5%; similarly, Duvick and Blasing (1981:1187) find an increase of approximately 3% relative to the same normals. However, the tree-ring data show find no increase at all relative to a 1941-to-1970 normal period.

The choice of which portion of the twentieth century constitutes a "modern normal" thus becomes critical: the 1931-to-1960 period includes the extremely dry decade of the 1930s, and it is therefore imperative to know whether or not conditions during this decade are particularly characteristic of the twentieth century. The drought of the 1930s is clearly represented in the Iowa tree-ring series, as are a number of other droughts. The decade of the 1930s, in fact, is only the fourth driest period represented in the 300-year series: in order of severity, droughts in Iowa date from 1816 to 1825, from 1735 to 1744, from 1696 to 1705, from 1931 to 1940, and from 1791 to 1800. Similarly, tree-ring studies by Stahle and Cleaveland (1988) find no significant changes in the intensity of droughts in north-central Texas between 1698 and 1980 and indicate that the 50 years "from 1931 to 1980 appear to have been generally representative of the last 283 years in Texas." Lawson (1974) also finds evidence in other tree-ring data for notable droughts in several portions of the Plains in the 1590s, 1680s, and 1850s.

The decade of the 1930s is therefore not distinctive of modern conditions but is paralleled by conditions extending as far back as there are data available in the Iowa region and elsewhere: droughts as severe or more severe than that of the 1930s are characteristic of the entire period that tree-ring data document. The apparent increase in precipitation on the Plains over the 1931-to-1960 period may thus reflect little more than the effects of a "normal" period that was too short to represent the actual range of climatic variation in the region rather than a significant long-term change in precipitation: the Iowa tree-ring data imply that "the abrupt changes in bias of actual 30-year precipitation averages in the twentieth century were not extreme . . . [and that] variations in 30-year means of estimated precipitation for the last 300 years reemphasize the limitations of 30-year normals for use in long-term hydrological planning" (Duvick and Blasing 1981:1188). Stahle and Cleaveland (1988:20–21) also note that 30-year intervals in their sequence often deviate significantly from long-term mean conditions and argue that "the 30-year 'standard normal' climate periods provide a reasonable measure of the current growing season moisture regime, but are less suitable estimates of long-term drought conditions." Given the variability in Plains climate on a decade-to-decade basis, then, a heavy reliance on Wahl and Lawson's (1970) 20-year records, assessed relative to the 1931 to 1960 "normal" period, seems unjustifiable in the face of conflicting evidence pertaining to longer periods of time.

Other empirical reconstructions based on tree rings do not show good evidence for significant increases in Little Ice Age precipitation. Fritts *et al.* (1979:33–35) reconstruct winter precipitation from 1602 to 1700, the period when Little Ice Age conditions were most severe, according to some interpretations (Lamb 1977:461–473), on the basis of tree-ring records throughout the western United States, including a number of tree-ring sequences from Wyoming and the Black Hills. Their results show only a slight decrease in precipitation in the Black Hills region between 1602 and 1650 and a slight increase in Kansas between 1651 and 1700, relative to the mean for the period from 1901 to 1970. Similarly, they reconstruct total annual precipitation on the Great Plains between 1602 and 1900 as within 0.7% to 3.7% of the 1901 to 1970 mean, with the greatest deviations (−2.1% and +3.7%) on the Southwestern and Northwestern

Plains (Fritts *et al.* 1981:158–159). The tree-ring analyses on which their reconstruction rests can be verified in part by comparing tree-ring-based inferences about weather conditions in specific years to recorded weather information for those years. Such a test for temperature and precipitation during the summer of 1849 proved to be extremely accurate, particularly for the Plains (Fritts *et al.* 1981:149–151).

These data suggest, at least, that the common assertion that annual precipitation on the Plains during the eighteenth and nineteenth centuries was substantially higher than the present is tenuously supported by the available data, and, at most, that it was, on the average, quite similar to that seen in modern times; Stahle and Cleaveland (1988:21), in fact, find a slight overall *decrease* in drought severity in Texas from the end of the seventeenth century to the present. The largest deviations reconstructed by Fritts and others (1981:158–159) are an increase of 3.7% on the Northwestern Plains and a decrease of 2.1% on the Southwestern Plains. The direction and geographic position of this latter decrease, although not its magnitude, agree with the theoretical predictions of Sanchez and Kutzbach (1974), and the increase similarly conforms with Reher and Frison's (1980:40) assertion that conditions on the northwestern Plains were wetter, although the contrast with modern conditions is relatively small: a 4% increase in present precipitation in Sundance, Wyoming, for example, implies a change from 430 to 447 mm (based on 1951 to 1980 precipitation data from the National Environmental Data Service; see Chapter 8); extrapolating roughly from Sims and Singh (1978b:579), this translates into an increase in forage production from about 1,900 to about 1,980 lb per acre.

The specific reconstruction of decreased precipitation between 1602 and 1650 in the Black Hills region is also supported by Reher and Frison's (1980:53–59) data on varve thicknesses from the Vore site in this region. Although these data are not calibrated to absolute amounts of precipitation, varve thickness increases as rainfall increases, and thicker varves therefore indicate wetter years. The plotted varve thicknesses for the period that corresponds approximately to the early 1600s appear in general to be below the mean of the sequence as a whole, although a few years show peaks in thickness and thus in precipitation during this time.

Average conditions, though, are only part of the overall pattern of climate; deviations from these conditions are also critical. Overall, a major characteristic of the Little Ice Age climate appears to have been an increase (relative to modern conditions) in the year-to-year variation in temperature and precipitation (Lamb 1981:302–305; but see Ingram *et al.* 1981:11–14, on problems with this conclusion), and this may have been characteristic of the Great Plains as well. Although the Vore site varves cannot be compared directly to modern data, the year-to-year variation in their thicknesses increases dramatically after the early 1500s (Reher and Frison 1980, Figure 35). Winter and summer temperatures over the entire

United States also appear to have been more variable than at present, and extreme winters were more common (Fritts *et al.* 1979:40–44).

Stockton and Meko (1983) also provide data on regional patterns of variation in precipitation. Plots of annual rainfall reconstructed from tree rings from four areas of the Plains show definite regional contrasts in the degree of year-to-year variation, with the reconstructions for the Southern Plains showing dramatically greater annual differences than reconstructions from any other areas.

To summarize, currently available data on climatic conditions on the Plains between 1450 and 1850 indicate that temperatures overall were cooler than at present, primarily because of colder winters. Precipitation appears to have been almost the same as at present, with a slight increase on the northwestern Plains and a slight decrease on the southwestern Plains. The seasonality of these minor changes is not indicated, but the likelihood noted before that there was little or no change in winter precipitation suggests that they refer to the rest of the year. Annual variation in temperature and precipitation may have been greater than at present. As in modern times, this variation appears to have been most extreme on the Southern Plains.

These patterns have predictable effects on forage conditions and hence on bison adaptations throughout the Plains. Even a small increase in annual precipitation on the Northern Plains coupled with a small decrease on the Southern Plains would tend to increase the differences in modern forage production noted between them in Chapter 5. The lack of significant change in summer temperatures implies that the relative proportions of warm-season grasses in different parts of the Plains were probably much the same as at present, but colder winters would have reduced grass growth during this season even below the extremely limited modern levels. Finally, annual fluctuations in these conditions, leading to parallel fluctuations in annual forage production and the timing of the growing season, appear to have been at least as great as at present. These fluctuations were substantially larger on the Southern than on the Northern Plains, as they are in modern times.

2. LITTLE ICE AGE BISON ADAPTATIONS

The general pattern described in Chapter 5 for the modern Plains grasslands appears also to have characterized the Little Ice Age grasslands. This pattern includes a clear northeast-to-southwest decline in the quality and quantity of forage produced, a north-to-south decrease in the proportion of cool-season grasses and therefore in the proportion of the year during which optimal forage conditions existed, and a much greater degree of year-to-year variation in forage conditions on the Southern than on the Northern Plains. The slight increase in precipitation on the Northwestern Plains and concomitant decrease in precipitation on the Southern Plains would have enhanced the first of these patterns.

A patchier distribution of rainfall on the Southern than on the Northern Plains would also have produced a patchier and less predictable distribution of actively growing and therefore more nutritious grasses. The closer relationship between growth patterns and local precipitation in warm-season grasses and the greater dominance of these grasses in the south would also have strengthened this pattern.

The likely effects of these differences on the bison are summarized in Table 6-1, which presents the relative differences in bison adaptations to be expected at the extremes of the conditions just summarized. As Chapter 5 shows, these conditions vary on the Plains along a roughly northeast–southwest axis from more productive grasslands with longer and more continuous periods of forage production and less year-to-year variation in forage conditions in the northeast to less productive grasslands with shorter and more erratic periods of forage production and greater year-to-year variation in forage conditions in the southwest.

Table 6-1. Predicted Variation in Bison Adaptations in Response to Differences in Environmental Conditions

Condition	Bison responses
Forage production	
Low	Population density low; herds small; herds move faster, farther, and more frequently; home ranges larger
High	Population density high; herds large; herds move more slowly, over shorter distances, and less often; home ranges smaller
Growing season	
Period of high production short	Longer period of seasonal undernutrition
Period of high production long	Shorter period of seasonal undernutrition
Growth erratic	Herd movements irregular; distribution of animals fine-grained
Growth continuous	Herd movements regular; distribution of animals coarse-grained
Forage distribution	
Patchy	More frequent and longer herd movements
Homogenous	Less frequent and shorter herd movements
Annual variation in forage conditions	
High	All of the above relatively variable from year to year
Low	All of the above relatively constant from year to year

Bison adaptations should then follow a similar geographic continuum, showing the differences summarized in Table 6-1.

Under ideal conditions, each of these different aspects of bison adaptations could be rigorously quantified, and different areas of the Plains could be readily compared, but given the difficulties in estimating factors such as regional carrying capacities even for modern environments, an attempt to do this on the basis of the data available at present would produce nothing more than a spuriously precise numerical smokescreen. Despite this problem, it is worth noting that the differences represented by the extremes of the continuum reconstructed here are likely to be great: forage production on the Northeastern Plains is roughly ten times as high as in the Pecos Valley, for example, and variation in annual precipitation is twice as high in the Texas Panhandle as it is in North Dakota. The range of variation in bison population densities, herd sizes, degrees of mobility, and other characteristics must have been correspondingly great.

Moving from the northeast to southwest across the Plains, there were probably progressively fewer bison more dispersed over the landscape in smaller and more mobile herds having progressively less regular patterns of movement within larger seasonal ranges. Seasonal changes in herd sizes would also have been less in the south and west because the areas with the lowest forage production would not have produced sufficient food to support extremely large herds at any time of the year. In addition, the degree of change in these conditions from year to year must have increased in the same direction, and seasonal undernutrition would also have persisted for more of the year among more southern herds. The major exception to this overall trend should have been in the Black Hills, where higher precipitation and cooler summer temperatures would have improved forage conditions over the surrounding area and herds would have been larger, more densely distributed, less mobile overall, and more regular in their movements (cf. Frison 1978:191–192).

Unfortunately, the predictions just discussed are difficult to test systematically because direct observation of free-roaming bison on the open Plains is simply impossible. The herds that live on the Plains grasslands all exist within restricted areas and are managed to one degree or another (ICWRU 1980; McHugh 1958; Shackleton 1968). Less restricted herds inhabit regions that are either near but are not actually on the Plains (Fuller 1960, 1961; Meagher 1973; Soper 1941) or are climatically extremely different from the Plains (Lott and Minta 1983).

There are two less direct means of assessing the predictions given. The first is by examining historical records of bison to see if the data these observations provide are consistent with theoretical expectations. Several problems with this kind of analysis were noted in Chapter 4, particularly the general absence of supporting environmental data in historical records and the disruptive effects of white expansion on the bison, particularly during the nineteenth century (Bamforth 1987 discusses these issues in detail). Historical data must therefore be used cautiously.

The second way to evaluate these predictions is by considering data on modern bison to see whether the processes identified in Chapter 4 as controlling ungulate adaptations in general also control bison adaptations in particular. If they do, there is good reason to accept the predictions just outlined. The following discussion combines these two sources of information and follows the structure of Chapter 4, discussing regional population densities, patterns of food selection, the existence and size of familiar areas and home ranges, and migration and aggregation patterns.

3. BISON POPULATION DENSITIES

As was discussed earlier, ungulate population densities are regulated by the amount of forage produced in a region. This factor is critical because the nutritional content (particularly the protein content) of most forage drops below the levels that ungulates require to maintain their body weight during the nongrowing season, and seasonal undernutrition is therefore common among these animals.

In the worst years, this undernutrition can cause starvation, but even in less extreme circumstances, it renders animals less resistant to disease, reduces fertility, and can so weaken many of the young born to undernourished mothers that they are unable to survive their first year of life. Speth (1983:121–131) has shown that forage conditions on the Plains show exactly the pattern of seasonal change that leads to this process of population regulation, and modern range managers are frequently urged to provide protein supplements for their livestock during the winter (Bokhari 1978b; Cogswell and Kamstra 1976; Rauzi *et al.* 1969; Willard and Schuster 1973).

Data on mortality patterns among modern bison are relatively few and lack much of the supporting information that would provide a rigorous test of the theory that bison populations are regulated by overall forage production. The most important missing information is on the nutritional status of deceased animals, which the theory predicts should be poor. Modern studies of bison have also not considered the effects of the drop in nutritional value of grasses in the winter. However, the available data are at least consistent with theoretical expectations. All authors who discuss the causes of bison mortality note that most animals die in the winter (Fuller 1961; Meagher 1973; Soper 1941; also see Frison 1978:9, 11), exactly the season when grasses are least nutritious and the detrimental effects of undernutrition should be manifest. Meagher (1973:73) states this most explicitly:

> Winterkill, probably from the combined effects of climatic stress, forage availability, and physiological condition of individual animals, was the main cause of observed mortality. . . . Death usually occurred after prolonged weakening, often in late winter. . . . A few animals died annually, but the number increased with severe winters.

In addition to this, Meagher (1973:56) notes that bison in Yellowstone Park have lower reproductive rates than bison elsewhere. She tentatively attributes this to worse winter conditions and hence worse winter nutrition in the park than in other areas.

Historic records also attest to the effects of winter forage conditions on the Plains bison. The effort that bison had to exert to obtain forage during the winter in at least some parts of the Plains is indicated by traces of blood around areas that the animals had cleared with their muzzles for grazing, caused by the hardness of the packed and sometimes icy snow (e.g., Roe 1970:201). Observers also commonly noted the poor condition of the bison at the end of the winter. Audubon and Bachman (1854:46) described bison at the end of a severe winter as "emaciated" and "wretched," and Henry's journals (Coues 1897:594) describe bison in April as "so weak that if they lie down they cannot rise" (also see Dodge 1884:283; Hornaday 1889:423). Archaeologically, Speth's (1983) analysis of the butchering and bone-disposal patterns at the Garnsey site in eastern New Mexico where a series of early spring kills took place also suggests strongly that bison were malnourished over the winter.

It is therefore highly likely that the sizes of bison populations on the Plains were regulated by the availability of food during the winter. This implies that overall bison population densities followed the pattern of forage production shown earlier in Figure 5-5, decreasing steadily and substantially from northeast to southwest except in the vicinity of the Black Hills, which should have supported substantial herds. The likely lower nutritional content of the grasses on the Southern Plains also implies that bison numbers probably declined somewhat faster than would be predicted by forage production alone because grazing animals need to consume more of these grasses to obtain adequate nutrition. The areas of lowest forage production are currently found in the Pecos River Valley, and it is therefore likely that bison numbers here were also lowest.

Although the earliest historical records for the Plains are from this last region, they provide regretably sketchy information on either overall bison densities or specific herd sizes. However, the limited evidence in these documents is consistent with the expected overall distribution of bison on the Plains.

Neither of the two major Spanish expeditions up the Pecos River (those of Antonio de Espejo in July 1583, and Castano de Sosa in November and December 1590) found any bison, although the Espejo expedition found bison hoofprints and bones, and de Sosa records many deer and a game corral (Hammond and Rey 1966:207–208, 259–265). Similarly, expeditions north into Colorado along the eastern edge of the Rockies (led by Diego de Vargas in October 1696, Juan de Ulibarri in July 1706, and Juan Paez Hurtado in August 1715) mention no observations of bison, although they do note large numbers of deer and prairie chickens and otherwise describe the country in some detail (Thomas 1935:53–59, 59–76, 94–98).

Francisco Vasquez de Coronado's expedition in the spring of 1541, the earliest in the region, encountered no bison until it had traveled 8 days to the east of the Pecos River; the records of the journey subsequently note "incredible" numbers as the expedition continued on to the northeast (Hammond and Rey 1940:235–237). The lone Indian survivor of an illegal expedition led northeast from Pecos by Francisco Leyva de Bonilla and Antonio Gutierrez de Humana in 1593 stated that "the farther inland they went, the larger was the number of buffalo they saw" and that the Spanish were amazed at the numbers of the bison they found in the regions beyond Quivira (Hammond and Rey 1966:324).

The expedition of Antonio de Valverde in October 1719, followed the route taken by Vargas, Ulibarri, and Hurtado into Colorado and then turned east past the point where these other expeditions turned back. Passing near Pueblo, Colorado, they saw a herd of 200 bison; continuing eastward down the Arkansas River, they first saw "many herds," then "many herds in all directions," and finally "great herds of bison so that in the distance they looked like rolling hills". Valverde estimated this last aggregate as at least 8,000 animals (Thomas 1935:110–132). Similarly, two earlier expeditions led by Onate (in the fall of 1599 and June of 1601) encountered no bison until they had traveled for some distance east of the Pecos River into the Canadian River Valley (Bolton 1908:223–226, 250–255). As Onate continued farther east on the second journey, he notes that the land continuously became more verdant and describes the bison in it as "innumerable" (Bolton 1908:256–257).

Farther to the south, Juan Domingues de Mendoza traveled east from El Paso in December 1683 and returned in May 1684. Passing through Trans-Pecos Texas and then northeast into the region near Ballinger, he first noted bison tracks and then encountered and killed three bison bulls near Fort Stockton, alleviating great hunger in his camp. Continuing to the northeast, the expedition killed steadily increasing numbers of bison and finally camped with many Indians for 6 weeks, during which time they killed 4,000 bison. Traveling toward the southwest on the return trip, Mendoza then recorded steadily decreasing numbers of kills per day. He also described the area in which the last five of these kills were made as noticeably drier than the region through which they had traveled previously (Bolton 1908:320–343).

Historical observations of bison must always be interpreted cautiously because they generally lack important information on local environmental conditions affecting the herds and because the more recent records document a period of severe disruption for the Plains bison (Bamforth 1987). However, the Spanish records discussed here antedate this period of disruption, and the consistency of the pattern shown in observations made during different seasons and in different years implies that variability caused by local environmental conditions was not great enough to obscure fundamental regional patterns.

All of the Spanish data fit well with the expectation that bison densities increased across the Plains to the north and east and particularly well with the hypothesis that these densities were especially low in the Pecos Valley and low on the Southern and far Western Plains in general. The Spanish also seem to have noticed the relatively low bison densities on the Western Plains margins. Referring to Ute territory in Colorado, adjacent to the Rocky Mountains, Escalante (cited in Secoy 1953:29) states that "in the plains of the Yutas there are not many buffalo." Given the variability in climate in the region, there can be no doubt that there were fluctuations in bison numbers in any given local area from year to year. However, these early historical records seem to imply fairly clearly that noticeable, consistent regional differences in bison numbers did exist.

The apparently low density of bison on the Southern High Plains in particular is in direct opposition to the picture presented by ethnographic research. Colson (1954:14), for example, states that "the Comanche [on the Southern High Plains] lived in the best buffalo country," and Wallace and Hoebel (1952:34) state that "of buffalo they [the Comanche] had a plethora and were more richly supplied than the tribes further north," but none of these authors presents any evidence supporting these assertions. Comanche enthnography draws largely on historical records and on interviews with aged informants conducted in 1933 and 1945 (Wallace and Hoebel 1952:x–xi), and it seems reasonable to question whether either of these sources of information provides reliable data on bison ecology during the eighteenth and nineteenth centuries. It is possible that the slaughter of the bison on the Central and Eastern Plains in the late 1800s led the surviving animals to use the relatively inaccessible Southern High Plains as a refugium, increasing the number of bison in the region during this time. However, the data on the nutritional status of bison and seasonal forage conditions on the Plains are too clear to project this pattern to the periods before perhaps 1860. It is therefore likely that the traditional view of bison numbers on the Southern High Plains is incorrect, at least for periods before the late nineteenth century.

4. BISON FEEDING STRATEGIES

The bison is a ruminant like its relative, the African buffalo, and many other large ungulates. Adult males weigh approximately 700 to 900 kilograms, and adult females weigh approximately 350 to 500 kilograms (Meagher 1973:38–39; Wheat 1972:85–86), placing them in the upper range of ungulate size (compare

Jarman 1974; Meissner 1982:46–47). Following the discussion in Chapter 4, this implies that, although they can tolerate a relatively low-quality diet, they require fairly large absolute volumes of forage.

Several studies of the feeding strategies of bison on the Plains (ICWRU 1980; Meagher 1973:90–95; Peden 1972) show that they are almost exclusively grass eaters, although they will consume browse when other food is scarce, particularly during seasons when grasses are of least nutritional value (Martin *et al.* 1961), and they prefer sedges to grass in more northern regions (Reynolds *et al.* 1978). Like other unuglates, bison also tend to select the most nutritious forage available at any given time, consuming cool-season grasses when they are most nutritious during the spring and fall and switching to warm-season grasses during the summer when these species have their most active period of growth (Coppock *et al.* 1983b; ICWRU 1980; Peden 1972).

5. FAMILIAR AREAS AND HOME RANGES

Following Baker (1978), Chapter 4 noted that all individual animals move within a relatively restricted portion of the area inhabited by their species and that the same area is often inhabited from year to year. Similarly, the more restricted home ranges used by an animal during any given season are often much the same from year to year. In addition, nonoverlapping or minimally overlapping familiar areas or seasonal home ranges are likely to develop in regions where food resources are variable in location and where it is often necessary to move between feeding areas on the basis of decisions about where food is likely to be found (Baker 1978:399). If areas overlap, these decisions are more likely to be wrong than if they do not overlap, because animals in other groups may have already exploited and thus depleted the resources in the chosen area.

The Plains grasslands present a situation of this sort, in that forage production depends in large part on variable local rainfall patterns. As bison reduce the forage in one area, they must move elsewhere, and knowledge of good feeding areas in their region along with knowledge of where they had recently fed would clearly aid in deciding where to go. Studies of modern bison (McHugh 1958; Meagher 1973; Soper 1941) have uniformly indicated that fairly well-defined herds exploit distinct areas that remain essentially the same from year to year and that these herds move seasonally into specific portions of these areas. These areas often include a relatively small portion of the total area available to them (i.e., Soper 1941:380–384).

Reher (1974:123) also presents figures showing that the modern density of bison in Hayden Valley, in Yellowstone Park, is three times higher on their summer range than on their winter range, indicating that the herd exploiting this region tends to aggregate within a smaller area in the summer and is more dispersed over a larger area in the winter. This is the general pattern expected throughout the Plains, given that forage production is greatest in the summer and least in the winter, although the degree of aggregation should be less in areas that produce less forage overall. The total area exploited by the Yellowstone bison also appears to expand as populations increase and contract as populations decrease (Meagher 1973:29–37).

In other words, modern bison herds have readily identifiable familiar areas and seasonal home ranges whose size and location appear to depend on the forage needs of the population and the distribution of forage in the region they inhabit. There appears to be a low rate of exchange of members between these herds, as would be expected if knowledge of the specific territory the herd was exploiting was critical to survival. Meagher (1973:98–103) stresses the importance of thermal areas to bison in Yellowstone Park because the warmth of these areas reduces the snow cover on them, making it easier for the bison to forage. She suggests that in severe winters the amount of forage available in thermal areas may regulate population size. The locations of such areas can only be learned through experience, clearly indicating the importance of knowing a specific home range.

6. MIGRATION AND AGGREGATION PATTERNS

Bison migration and aggregation patterns have been the subject of most of the anthropological debate over bison ecology, and the essence of this debate is tied to the question of how predictable these patterns actually were. As Plains anthropologists have used the term, *predictability* seems to mean the degree to which herds of the same size returned to the same point on the ground at the same time of the year in successive years. Historical data that seem to show deviations from this ideal pattern or that seem to show few or no changes in herd sizes and distributions from season to season are often taken to indicate that the bison were "unpredictable" in this sense (Hanson 1984; McHugh 1972; Roe 1970).

The data discussed in Chapter 4, though, indicate that ungulate aggregation and migration patterns may be quite predictable without also being exactly the same every year. The problem is to recognize, first, the factors determining these patterns, such as the distribution of rainfall in a region, and, second, the ways in which animals respond to these factors. As Chapter 4 discusses, abundant data indicate that seasonal distributions of food and water are the key determinants of ungulate aggregation and migration patterns and that the effects of these distributions can be predicted with some confidence.

To discuss these topics, it is first necessary to outline the basic "seasonal round" of the bison, a topic that has been summarized extensively by various authors (Arthur 1975:39–60; Fuller 1961; McHugh 1958, 1972:179–205; Meagher 1973; Roe 1970:94–118; Soper 1941). For the immediate purposes of the present discussion, the key aspects of this round can be presented very briefly and relate primarily to reproductive behavior; more detail is added later.

Bison give birth in spring. Most births occur in April and May, although occasional calves are born throughout much of the year. The calving season begins somewhat later than this on the Northern Plains, where severe winter conditions persist later in the year, and somewhat earlier on the Southern Plains, where spring comes earlier.

In the calving season, as throughout most of the year, the bison travel in two relatively distinct types of groups: *cow/calf* groups (also referred to as "nursery herds" or "mixed" groups), composed of cows, newborn calves and yearlings, and sexually immature bulls, and *bull groups,* composed of sexually mature bulls and, less frequently, cows without calves. Cow/calf groups in modern herds typically range in size from 20 or 30 to several hundred whereas bull groups range from one or two to a maximum of about 30, and herds of both types are larger in open grasslands than in more forested areas (Fuller 1960; Hanson 1984; Shackleton 1968; cf. Mloszewski 1983:58–59; Sinclair 1977:120). Social organization within these two types of herds seems to differ slightly: although the females maintain a stable dominance hierarchy that is apparently based on age (Rutberg 1983), the males seem to have a less stable structure based on pairwise interactions between individuals in which status positions are often reversed though combat (Lott 1979).

Several months after calving, the female bison come into heat, and social groups composed of both sexes form for the rut, which occurs primarily during July and August. Access to estrous females during this period is governed by the somewhat flexible dominance relations between males: individuals with overall higher status obtained by more frequent victories in fights breed more often than individuals with lower status (Lott 1979). Following the rut, the cow/calf and bull groups re-form for the winter.

This basic reproductive and social cycle is recognized by most anthropologists. Arthur (1975:52) tends to emphasize anecdotal historic records of out-of-season births, but archaeological evidence shows a clear clustering of births within a fairly restricted portion of the year, presumably the spring (Reher and Frison 1980), and modern studies of bison support this pattern. Dispute over bison herding behavior within this basic framework has concentrated on the size of the herds present at different times of the year and the movements of these herds over the landscape.

The traditional view of bison aggregation patterns derived from this cycle sees the herds being segregated for most of the year into smaller cow/calf and bull groups, forming larger groups only during the rut (Dodge 1884; Frison 1974;

Hornaday 1889; Oliver 1962). In opposition to this, Arthur (1975) has argued that the herds were larger in the winter than during the rest of the year. He bases this argument on historical records of large herds of bison on the far Northern Plains and on modern data on bison collected by Soper (1941), which seem to indicate that winter herds were larger than other herds. A recent reanalysis of Soper's data by Hanson (1984), though, shows clearly that this is not the case and that herds are largest during the rut. Other modern herds show this same pattern (ICWRU 1980; McHugh 1958; Shackleton 1968).

Maximum herd sizes should also have been lower in areas with lower forage production, as Chapter 4 discusses. Some limited historical data tentatively support this prediction. Roe (1970:345–366) presents records of herd sizes throughout much of the Plains during the late 1700s and 1800s. As Bamforth (1987) discusses, these data show a pattern of increasing herd sizes through much of the nineteenth century, probably as a result of increasingly heavy human predation. Records prior to 1820 document a period before excessive predation began (Hornaday 1889), and they may therefore provide reasonable estimates of precontact herd sizes.

Roe's few pre-1820 records come from the Central and Northern plains and generally note herds of several thousand animals, ranging up to perhaps 10,000 or 20,000 animals during the rut. In contrast, Spanish records of herd sizes on the Southern Plains, all from the sixteenth and seventeenth centuries, note herds of no more than a few hundred animals (Table 6-2). Given lower forage produc-

Table 6-2. Historical Records of Bison Herd Sizes on the Great Plains before 1820[a]

Date	Herd size	Region
1581	500+	Southern Plains
1581	200–300	Southern Plains
1694	500+	Southern Plains
1719	200	Southern/Central Plains
1719	8,000 (multiple herds)	Southern/Central Plains
1801	1,000 and 5,000	Northern Plains
1804	3,000	Central Plains
1806	3,000	Southern Plains
1806	9,000	Central Plains
1806	10,000	Northern Plains
1806	20,000	Northern Plains
1811	600–800	Northern Plains
1811	10,000	Northern Plains
1819–1820	Several hundred	Northern Plains
1820	10,000	Central Plains

[a]From Hammond and Rey 1966:91, 131, 136; Roe 1970:345–350; Thomas 1935: 123, 129. The regions listed in this table refer to the areas defined in Chapter 1.

tion on the Southern Plains than elsewhere, smaller herds are expected there. The uniformity of herd sizes elsewhere on the Plains that the data in Table 6-2 tentatively suggest may indicate that under normal conditions there may have been an upper limit set by social factors, similar to that noted for the African buffalo by Mloszewski (1983:49). Herds larger than several thousand may not have been able to maintain a consistent dominance hierarchy and so may have tended to splinter.

The most extreme view of the regularity and scale of bison movements was popular during the nineteenth century. It held that the herds moved north onto the Canadian plains in the summer and then south into West Texas for the winter (see Hornaday 1889 in particular). Roe (1970:521–600) has shown exhaustively that such a regular pattern of movement is not only physically and ecologically impossible but is also contradicted by abundant historical documentation. Having refuted this notion beyond any reasonable doubt, Roe stresses the unpredictability and irregularity of bison migrations, although he does allow for some degree of constancy in the movements of the Canadian herds into the relatively sheltered aspen parklands on the edges of the Plains in winter and back onto the open plains in the spring (1970:570–574). The regularity of migrations in this area has also been discussed by Arthur (1975), Moodie and Ray (1976), Syms (1977), and Morgan (1980). A strong emphasis on a high degree of unpredictability in bison movements is also apparent in Hanson (1984) and McHugh (1972).

Moodie and Ray (1976) have noted the conflict between the evidence for the seeming unpredictability of bison migrations in most areas and the evidence for regular seasonal migrations on the Canadian Plains, and their analysis suggests that a closer examination of the data on the Plains as a whole would be worthwhile. Their data show that, although the normal pattern of bison movement in the north was into the sheltered aspen parklands in winter and onto the open plains thereafter, there was a relatively high degree of documented variation in the timing and extent to which this occurred. They argue that this variation was a predictable response to known forces such as prairie fires and winter conditions that were too mild to drive the bison off the open grasslands.

The documented "typical" migration of bison on the Canadian Plains is consistent with the expectations presented in Chapter 4. Rather than remaining on the open plains in severe winters, the herds apparently traveled to more sheltered adjacent areas where forage, possibly in the form of browse rather than just grasses (cf. Latady 1985; Martin *et al.* 1961) and protection from high winds and extreme cold could be found. This would inevitably have concentrated the animals in a smaller area than they exploited during the rest of the year, creating locally relatively large aggregations such as those documented by Arthur (1975). Such migrations would have been unnecessary in milder winters when forage would not have been deeply buried by snow and cold and wind would have been less severe. The data compiled by Moodie and Ray (1976) indicate that, under these conditions, the migration did not occur. In contrast, the typical pattern of seasonal aggregation for the rut and dispersion into small herds for the winter noted in historic records south of Canada and for modern bison is exactly the pattern expected in a seasonal grassland environment with less severe winters on the average than those experienced by bison on the Canadian Plains.

7. SUMMARY

This discussion relies on modern studies of bison ecology and, cautiously, on historical records of bison herd sizes and locations to show that the generalizations about the determinants of ungulate adaptations presented in Chapter 4 are relevant to the Plains bison. Although not all of the predictions of these generalizations can be tested rigorously, the available information agrees well with them. This is not surprising considering the relatively close evolutionary relationship between bison and one of the best-studied species discussed in Chapter 4—the African buffalo. The basic structural similarity of the Plains environment to that of the East African grasslands also suggests that ungulate adaptations in the two regions should share a similar set of determinants.

Available information indicates that bison populations are regulated by their food supply. This therefore implies that under natural conditions bison densities were lower in areas with lower forage production than in areas with higher forage production, and available data indicate that there was indeed a northeast-to-southwest decrease in bison numbers across the Plains. Bison are also able to select more nutritious food than that available in the environment as a whole, as are other ungulates.

Use of familiar areas and home ranges also seems to parallel the pattern seen in other species, with evidence seen in modern bison for the existence of relatively discrete herds within fairly clearly defined areas. These herds regularly exploit specific portions of these areas at different times of the year. The size of familiar areas and seasonal ranges seems to depend on the amount of food and water present, being larger when food is less abundant and smaller when it is more abundant. Known patterns of migration and aggregation on the Plains also reflect the same climatic and other determinants important for other species.

The available data thus uniformly support the predictions made earlier in this chapter: observations of modern bison indicate that population densities and patterns of mobility and aggregation are controlled by the processes discussed in Chapter 4 and historical observations of bison are consistent with the specific predictions about regional bison populations, herd sizes, and seasonal migration patterns.

Chapter 7

Recent Population Movements on the Great Plains

Chapter 2 proposes a relationship between a society's complexity and the specific environment that that society exploits, but the specific locations of the recent occupants of the Great Plains changed dramatically during the centuries just before and after white contact. It is therefore necessary to consider these changes in studying the relationship between the eighteenth- and nineteenth-century Plains grasslands and the people living in them. This chapter discusses the pattern of hunter–gatherer migration across the Plains between A.D. 1650 and 1850, summarizing the tribal movements and the accepted explanations for them and identifying several points for which these explanations do not account. Chapter 8 tests the hypothesis proposed in Chapter 2 against data on the final distribution of the Plains tribes and then considers the migrations summarized here in light of the results of this test.

1. TRIBAL DISTRIBUTIONS FROM A.D. 1650 TO 1850

Although the recent migrations of many tribes onto the Plains are surely linked to European expansion, similar migrations occurred long before this expansion had any substantial effects on native culture. Unfortunately, it is difficult to reconstruct the locations of many tribes during the earliest centuries of white contact because of the limited information recorded by many of the first European explorers and because of the varying names applied to different native groups by these explorers. Some aspects of the distribution of hunting tribes before approximately A.D. 1650 are known, but many are not. At about this date, a relatively complete picture of these distributions can be reconstructed, and later periods, particularly after approximately 1700, are progressively better understood.

In considering tribal "territories" on the Plains, it is important to remember the high degree of mobility of the known bison-hunting groups in the region. The introduction of the horse obviously increased this mobility substantially, but even the pedestrian groups covered large territories and periodically traveled far outside of their usual lands to trade, hunt, and wage war. Boundaries indicated in this section should therefore be taken as indicating approximately the areas within which a given group habitually moved rather than as being clear territorial markers. Despite this ambiguity, native descriptions of tribal territories often note quite specific landmarks as boundaries, implying that territories were recognized even if they were not always respected.

The earliest European contacts with any tribes on the Plains were recorded by the Spanish under Coronado in 1541 (Hammond and Rey 1940), who traveled from Pecos Pueblo, through northeastern New Mexico, across the northern Texas Panhandle, and into Oklahoma and Kansas (Schroeder 1974). Although it is difficult to link all of the named groups that Coronado encountered on the High Plains with tribes known in more detail from later times, at least some of them were certainly Apache. Linguistic evidence indicates that the Plains Apache separated from their northern Athapascan ancestors substantially before white contact, possibly as early as A.D. 600 (Hoijer 1956; Perry 1980), and information gathered by the Spanish from the Pueblo Indians of New Mexico indicates that they arrived on the Southern High Plains in approximately A.D. 1525 (Gunnerson 1974; Hyde 1959), although archaeological data discussed by Spielmann (1983) suggest that they may have arrived as early as A.D. 1450.

Following this initial report, there is no documentary information on the Plains Apache until the late 1600s (Schroeder 1974:315). At this time, Apachean groups apparently occupied the Western Plains from southeastern Wyoming south through eastern Colorado and western Kansas into northwestern Texas (Gunnerson 1974; Hyde 1959:3–51; Schroeder 1974; Secoy 1951, 1953). Early Spanish descriptions suggest that the first Apache occupants of the region were full-time hunter–gatherers, but these later groups were part-time horiculturalists, spending the planting and harvesting seasons in small "rancherías" dispersed throughout their territory and living the rest of the year as nomadic bison hunters.

Documentary information on hunting groups on the rest of the Plains in the sixteenth and most of the seventeenth centuries is generally lacking. However, oral traditions collected from several tribes on the Northern and Northwestern Plains indicate general tribal locations, and some inferences can be drawn from later locations. The far northwestern Canadian Plains from Edmonton to McLeod were occupied by the Blackfoot throughout these centuries (Wissler 1910:15–

19), and the more eastern Canadian Plains were apparently occupied by the Assiniboine (Ray 1974:4–23).

During the late 1600s, ethnohistoric evidence indicates that the Assiniboine were fighting with the Dakota in northern Minnesota, with the Blackfoot, and with the Gros Ventre, who lived southwest of the Assiniboine (Ray 1974:14; also see Brink 1986). This last evidence is interesting because the seventeenth- and eighteenth-century locations of the Gros Ventre and the Arapahoe, from whom the Gros Ventre separated at some unknown time in the recent past, have been disputed (Trenholm 1970:8–12 summarizes this dispute). Early theories suggested that both groups moved out of Minnesota at the end of the seventeenth century. However, more recent analyses find no support for this hypothesis and place both groups on the Plains at a relatively early, although somewhat uncertain, date. Hewes (1948) suggests that the Gros Ventre and the Arapahoe split in the late 1600s in western North Dakota, a hypothesis that agrees with Ray's ethnohistoric data and further implies that the Arapahoe were on the Plains north of the Black Hills at this time.

The greatest problems in reconstructing tribal locations in the late 1600s are in the areas south and west of these groups. Only in western Montana are the locations and approximate limits of tribal territories fairly certain: Both Blackfoot (Wissler 1910:17) and Flathead (Teit 1930:304–305) traditions indicate that this area was occupied by the Kutenai and the Flathead, known in later times as Plateau groups occupying the Rocky Mountains and venturing into the Plains only seasonally. The Flathead also reported that at this time they had no knowledge of the Crow, Sarsi, Arapahoe, Gros Ventre, Cheyenne, or Assiniboine.

South of the Blackfoot, east of the Flathead and Kutenai, west of the Gros Ventre/Arapahoe and the horticultural groups along the Missouri River, and north of the Apache, there are three major tribes whose locations are somewhat problematic: the Shoshone, the Crow, and the Kiowa/Kiowa-Apache. By 1700, the Kiowa had made an alliance with the Crow and apparently controlled the area from the eastern Black Hills north along the Little Missouri River, with the Kiowa-Apache just south of them (Hyde 1959:137–139; Mayhall 1962:11). Archaeological evidence of Crow occupation at the Hagen site in this general region by this time seems clear (Mulloy 1942), and it is likely that the Kiowa were in this vicinity some time between A.D. 1650 and 1700.

The dates for the initial appearance of the Crow as a distinct tribe on the Plains are uncertain. The Crow were originally a part of the horticultural Hidatsa along the Missouri River, as attested by Crow and Hidatsa tradition and by linguistic studies (summarized by Wood and Downer 1977). The separation of these two groups has been placed at times varying from 14 B.C., on the basis of glottochronology, A.D. 1775, on the basis of ethnohistoric evidence (Wood and Downer 1977:83). Pointing out that linguistic divergence does not necessarily imply physical separation, Wood and Downer (1977) suggest that the Crow/Hidatsa division was probably a gradual process that may have begun in the mid-1600s and was completed during the period to which the ethnohistoric data refer.

The major problem with this theory, as Wood and Downer acknowledge, is the existence of radiocarbon dates as early as A.D. 1420 on sites in Wyoming bearing pottery thought to be of Crow manufacture (Frison 1976). This pottery is distributed through northern Wyoming and adjoining portions of Montana and South Dakota but has been dated at only a very few locations. Frison (1976) takes these dates as evidence for Crow occupation of this area possibly as early as the sixteenth century and extending through the eighteenth century. More recently, Keyser and Davis (1982) have found this type of ceramic in a site in southeastern Montana that is reliably dated between A.D. 1000 and 1100, substantially too early for any reasonable date on the Crow–Hidatsa separation. They propose that much of the apparently "Crow" material represents an indigenous but ethnically unidentifiable pottery-making society that interacted with groups along the Missouri River and suggest that this society was replaced by the Crow after A.D. 1400. Regardless of which of these explanations is correct, it seems clear that the Crow occupied the Bighorn Mountains in northeastern Wyoming and adjacent areas of Montana and the Dakotas by the end of the seventeenth century.

The last major group on the Northwestern Plains at this time was composed of Shoshonean or Numic speakers who had migrated out of the Great Basin, through the mountains of Utah and Wyoming, and onto the Plains some time after the fifteenth century; these groups were referred to historically as the Snake (Hultkranz 1968; Hyde 1959:117–145; Shimkin 1941; Wright 1978). Despite the undoubted presence of these groups in the region, there are problems in determining the limits of their territory prior to the introduction of the horse.

Some authors (Reher and Frison 1980:33; Secoy 1953:33; Wright 1978) assert or imply that the Shoshone occupied the eastern Rocky Mountains and ajacent arid basins of Wyoming until they obtained Spanish horses by about A.D. 1700, after which they expanded throughout the Northwestern Plains into Canada. However, the empirical basis for this assertion is obscure, and there are data (summarized by Sutton 1986) that do not conform to it.

An ethnohistoric account (Secoy 1953:34–35) of a battle between the Shoshone and the Blackfoot in approximately 1725, before horses or guns were used in combat, places the encounter on the Plains near the Eagle Hills in southern Saskatchewan. This battle is often characterized as the "first" Blackfoot–Shoshone battle (Secoy 1953:34), a characterization that may be the basis for the late date that is traditionally accepted for the Shoshone expansion (Sutton 1986). However, this account explicitly states that conflict between these two tribes was common in this region at this time, suggesting that it may have existed for some time and opening the possibility that the Shoshone expanded as pedestrian rather than mounted nomads.

This possibility is supported by archaeological data. Keyser (1975) has argued that the shield-bearing warrior motif in Northwestern Plains rock art is of Shoshonean origin, and Wright (1978) similarly points out that the distribution of this motif corresponds exactly to the apparent route of Shoshonean migration out of the Great Basin. The shield-bearing warrior is distributed on rock art over much of the Northwestern Plains (Keyser 1975, Figure 2) and is a major element in the rock art at Writing-On-Stone along the Milk River in southern Alberta (Keyser 1977, 1979).

Shoshonean-style warriors are shown in two contexts at Writing-On-Stone. The first context is in apparently prehistoric scenes depicting battles between two groups of pedestrian combatants, using only aboriginal weapons. The second depicts warriors mounted on crudely drawn horses attacking other warriors on foot, with these latter groups sometimes carrying guns. This pattern corresponds to the early Historic Period in this region, which is discussed in more detail later. Fully Historic paintings showing mounted combatants bearing the full complement of recent Plains war equipment are also present at Writing-On-Stone but are executed in a style that differs from that of the Shoshone and corresponds exactly to that in Historic paintings of the Blackfoot, Cree, Gros Ventre, and Assiniboine (Keyer 1977:55–57, 1979:43–45). Mulloy (1958:119–120) similarly dates art that is executed in a Shoshonean style to the late Prehistoric Period at Pictograph Cave, near Billings, Montana, on the basis of design content, superposition, and degree of weathering, and notes that the shield-bearing warrior is unknown in the art of the historic Plains tribes.

The likely antiquity of Blackfoot–Shoshone hostilities suggested here and the presence of Shoshone artists along Milk River during the late prehistoric period suggest very strongly that the Shoshone occupied most of the Northwestern Plains from Wyoming to the Canadian border before the introduction of the horse. This distribution corresponds almost exactly to the Shoshone territory described by the Flathead (Teit 1930:304–305) for the prehorse (late Prehistoric) period and proposed as a possibility by Hewes (1948) and, more definitely, Keyser (1975), although the exact date at which the Shoshone reached these boundaries is uncertain. Ethnohistoric evidence (Secoy 1953) indicates that they received horses shortly after A.D. 1700, and their arrival on the Milk River must therefore predate that time. Brink (1986) also summarizes ethnohistoric evidence that the "Snake" were on the Plains of southern Alberta before 1700.

Besides the Shoshone, two other distinct groups of Numic speakers are known on the High Plains at various times—the Ute and the Comanche. The Shoshone and the Comanche were one people when they first arrived on the Plains, separating from one another after the period discussed here, as is summarized later. The Ute appear to have expanded out of the Rocky Mountains onto the westernmost portions of the Plains in Colorado at about the time that the Shoshone arrived in Wyoming (Hyde 1959).

Figure 7-1 shows the approximate locations of these various groups in 1700. Despite the problems discussed before, this overall distribution appears to be consistent with the presently available evidence, although Brink (1986, Figure 2) locates the northern boundary of Snake, or Shoshone, territory and the southern boundaries of Gros Ventre and Blackfoot territory somewhat farther north at this time. Better data from later periods indicate that this distribution changed rapidly and dramatically over the next 150 years. Tribal movements during this succeeding period are relatively well-known and can be summarized briefly. The major population movements after A.D. 1700 involve, first, the southward expansion of

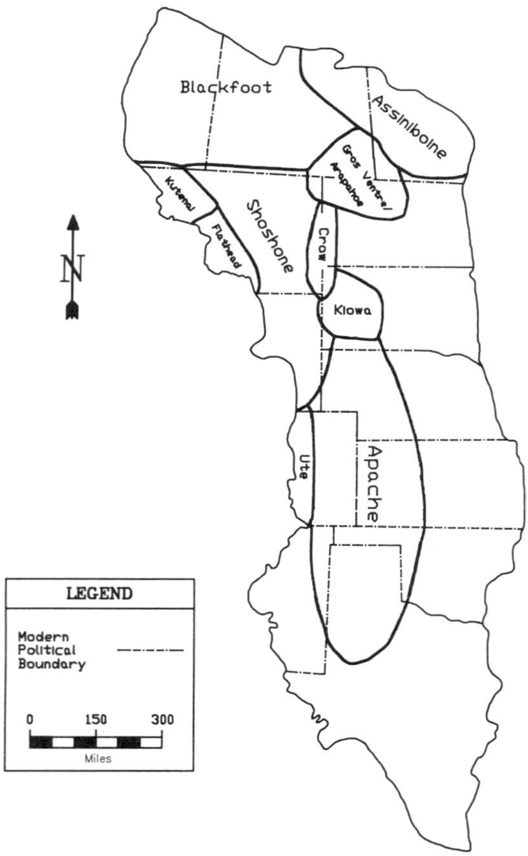

Figure 7-1. Tribal distributions on the Great Plains ca. A.D. 1700.

Numic speakers out of the Northwestern Plains and, second, the generally southwestward migrations of many other groups on the Plains, particularly of Siouan and Algonkian speakers from the Northeastern Plains periphery. The following summary is drawn from Hewes (1948), the various tribal entries in Hodge (1907, 1910), Hyde (1937, 1959, 1968), Ray (1974), Reher and Frison (1980:29–34), and Secoy (1953).

Although the northern limits of the Shoshonean expansion appear to have been reached before the appearance of the horse on the Northern Plains, the southern limits were not reached until considerably later. Between A.D. 1700 and 1725, Shoshonean groups drove the horticultural Apache out of the region from the Black Hills south to the Arkansas River. These southern Shoshonean groups after approximately A.D. 1726 are referred to as the Comanche, taken from the Ute word *Komantcia,* or enemy (Wallace and Hoebel 1952:4–5). Although the Comanche and the Ute raided and traveled as far south as the Spanish settlements in New Mexico during this time, they made no attempts to occupy the lands in this region permanently until later. Between A.D. 1725 and 1750, this southward advance halted, only to begin again in about A.D. 1750. By A.D. 1775, the Comanche had completely driven the Apache from the Plains, and occupied the Southern High Plains as far as central Texas and the Rio Grande River.

At approximately the same time, conflict on the northeastern edge of the Plains began to displace many groups toward the south and west. The Cheyenne, who at this time were horticulturalists, moved from southwestern Minnesota to the Sheyenne River in South Dakota by 1725. Constant pressure from the Dakota and other tribes drove them to establish new villages to the west on the Missouri River and, by 1780, finally to abandon agriculture completely and move into the Black Hills. By 1800, they had also been driven out of this region into northeastern Colorado and southeastern Wyoming. At this same time, particularly after approximately 1750, the Crow and the Blackfeet began to drive the northern Shoshone out of Montana and Wyoming, finally pushing them into the Rocky Mountains by 1800. The most northern Blackfoot were joined some time during the mid-1700s by the Sarsi, an Athapaskan group moving out of the adjacent woodlands.

The Teton Dakota followed the Cheyenne out of Minnesota, crossing the Red River into South Dakota by 1725, reaching the Black Hills just after 1800, and moving into the Big Horn Mountains in Wyoming by 1825. During the 1700s, the Assiniboine also gradually shifted their main territory west from southern Manitoba and northern Minnesota into the parkland belt and adjacent Plains in Alberta, pushing the Gros Ventre and the Arapahoe in front of them. Although the Gros Ventre remained in the north, the Arapahoe moved south into the Black Hills region before 1800, forming a close association with the Cheyenne. The movements of these new groups into the area near the Black Hills around 1750 forced the resident Kiowa and their allies, the Kiowa-Apache, the

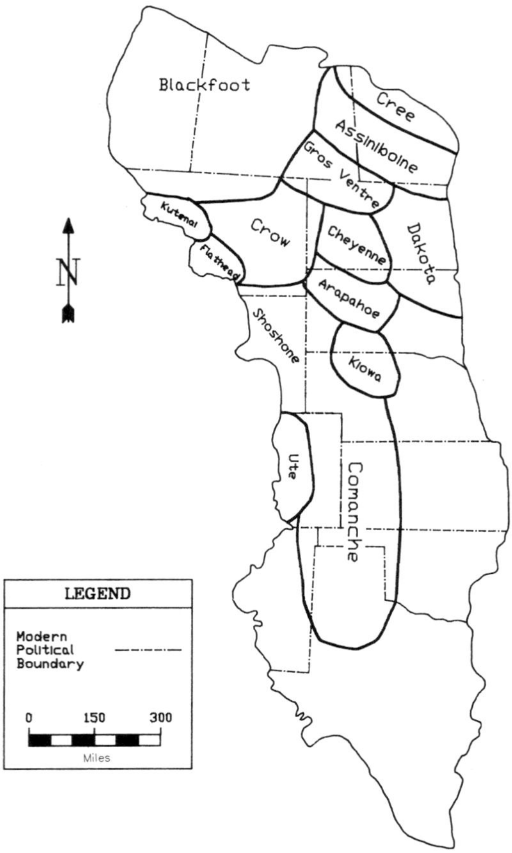

Figure 7-2. Tribal distributions on the Great Plains ca. A.D. 1775.

last Athapaskan speaking group on the Plains, to the south into Colorado and Nebraska before 1800.

These general movements continued into the mid-1800s, ending with the distribution of tribes recorded by modern ethnographers. This distribution placed the Comanche in the far southwest, primarily in Texas, with the Kiowa and Kiowa-Apache just north and east of them in Oklahoma. In the 1820s, the Cheyenne divided into northern and southern groups, with the northern groups moving into the area south of the Black Hills on the upper Platte River and the southern group moving to southeastern Colorado along upper Arkansas River; the Arapaho occupied the region between these two groups, and the Ute held only the very western Plains margin in Colorado. To the north, the Crow were in much of Wyoming and adjacent parts of Montana with the Blackfoot and the Sarsi north of them from Montana into Saskatchewan. The Gros Ventre were in

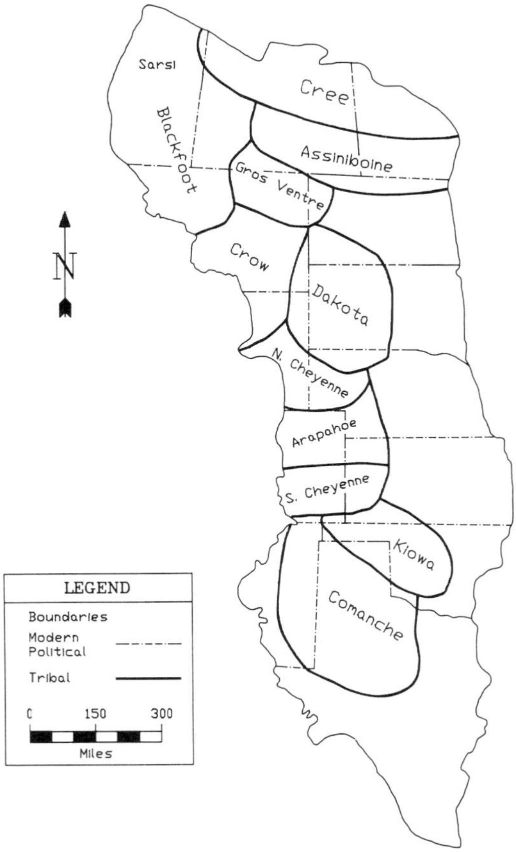

Figure 7-3. Tribal distributions on the Great Plains ca. A.D. 1850.

northeastern Montana and southern Alberta. The Teton Dakota held the Black Hills, southwestern North Dakota and much of western South Dakota and northwestern Nebraska, with the Assiniboine to the north and the Cree beyond them. Figures 7-2 and 7-3 show the approximate locations of the various groups on the Western Plains in 1775 and 1850.

2. HISTORICAL FORCES AND RECENT MIGRATIONS ON THE GREAT PLAINS

The causes of the earliest migrations on the Plains, particularly those of the Numic and Athapaskan groups before European contact, are poorly understood. However, migrations after this time are linked closely to changes in the ability of the various tribes to obtain European horses and firearms. The spread of these two imports across the Plains and the effects of this spread on the distributions of the tribes there have been documented in detail by several authors (Ewers 1955; Haines 1938a, 1938b; Hyde 1959; Ray 1974), particularly Secoy (1953). The following summary is drawn from this work.

Differences in tribal access to horses and firearms grew out of the differing policies of the European colonial powers toward trade. The Spanish in New Mexico and Texas were forbidden to trade guns or ammunition to the native tribes, although they could trade other goods, including metal knives and other tools. However, the Spanish had large numbers of horses, many of which the Indians obtained through purchase and theft. In contrast, the French in Canada and New Orleans and the English on Hudson's Bay had few horses but were quite willing to trade in firearms and ammunition. These contrasting policies created distinct gradients in the distributions of horses and guns across the Plains, with the numbers of horses increasing toward the southwest and the numbers of guns increasing toward the northeast.

The horse was distributed widely across the Plains earlier than was the gun. The earliest Plains groups to obtain large numbers of horses were the Apache, who lived closest to the Spanish settlements in New Mexico. The Plains Apache obtained substantial numbers of horses some time after 1650 and rapidly developed a style of mounted warfare that gave them a dramatic military advantage over their neighbors, an advantage that was enhanced by Apache use of leather armor to protect both the horses and the rider from arrows.

Horses then spread to the Caddoan tribes to the east by 1700 and were obtained in large numbers after 1690 by the Ute and at about 1700 by the southern Shoshone (Comanche). Large mounted bands of Comanche and Ute then turned on the small, seasonally occupied Apache horticultural villages one by one and drove their occupants south of the Arkansas or west across the southern Rocky Mountains. At the same time, these groups rapidly distributed horses to their northern relatives, and the Shoshone in Montana and Wyoming were raiding the Crow and the Blackfoot on horseback by 1730. Other northern tribes acquired increasing numbers of horses during this time, mainly by stealing them from the Shoshone, and, by 1750, the Blackfoot, for example, were essentially fully mounted.

Firearms, in contrast, were first available to the tribes on the eastern periphery of the Plains and diffused slowly to the west. In the north, the Cree and Assiniboine were first extensively armed by the English around Hudson's Bay after 1670 and generally maintained access to guns and ammunition as English and French influence in the region waxed and waned. In the 1730s, the tribes allied with the Crow and Blackfoot against the Shoshone were using a few guns in the fighting on the Northwestern Plains, and Secoy (1953:51) refers to the mid-eighteenth century as the "Few Guns Period." Lacking access to French and

British traders and unable to get firearms from the Spanish, the Shoshone began to lose their military advantage. After the British obtained permanent control of the northern fur trade in 1763, the number of guns available to the Shoshone's enemies increased substantially, and the Shoshone were progressively driven from the Plains.

To the south, horses were relatively widespread before many guns were available to the native tribes. The Caddoan groups in the Eastern Plains began to obtain guns from the French in exchange for Indian slaves after 1700, and Secoy (1953:79) dates the "Few Guns Period" on the Southern Plains between this date and approximately 1720, by which time the Pawnee on the Platte River in Nebraska were sufficiently well-armed to destroy a Spanish expedition from Santa Fe led by Villasur (Thomas 1935:33–39). A few guns moved from these eastern groups to the Apache at this time, but they were not numerous enough turn the tide against their Ute and Comanche attackers. In approximately 1740, the Comanche made an alliance with the Caddoan tribes in eastern Nebraska, Oklahoma, and Texas and thereby obtained a reliable source of guns for themselves. By 1750, Hyde (1959:107) states that every Comanche warrior coming to Taos to trade carried a French musket. The refusal of the Spanish to supply the Apache with guns even in the face of this situation essentially ensured the ultimate Comanche victory.

These data thus indicate that human migrations on the Plains during the Historic Period were strongly related to the relative military capabilities of the various tribes: as changing access to horses and firearms shifted the balance of power in a region, stronger groups advanced at the expense of weaker groups. In the classic discussion of this relationship, Secoy (1953) particularly links the advance of the Shoshone across the Northwestern Plains and of the Comanche into the Southern Plains to the greater access of these groups to horses and the subsequent defeat of the Shoshone in the north to their inability to obtain the firearms that their enemies had in abundance.

Despite the importance of these historical factors, three points indicate that they do not provide a complete explanation for the timing and directions of historic migrations on the Plains. The first of these is the evidence presented before indicating that the Shoshone expanded across Wyoming and Montana to the Canadian border in late prehistoric times, before the introduction of the horse. The second is the ability of such groups as the Kiowa and the Cheyenne to push the Comanche south of the Arkansas into the least productive, most unpredictable part of the Plains after 1750, despite the abundance of both horses and guns among the Comanche. The continued triumphs of the Dakota over mounted and well-armed rivals well into the 1800s are similarly difficult to account for only by reference to historical factors.

Finally, if access to horses and guns provided a sufficient explanation for the expansion of neighboring groups onto the Plains, such groups should have expanded whenever and wherever these innovations became available. Instead, peripheral groups expanded only from the Northeastern Plains margin, despite the presence among the Caddoan tribes to the south of Spanish horses by 1690 (Haines 1938b) and French firearms by the early 1700s (Secoy 1953:79–81), as early or earlier in both cases than among any of the groups on or adjacent to the Northern Plains. The Apache in the south had an early military advantage because of their greater access to horses at an early date that undoubtedly helped to hold their territory, but this advantage disappeared after horses spread to the east and combined with even moderate numbers of French guns. Despite this advantage, the Caddoan tribes on the edges of the Plains expanded their territory only within areas where they could farm, taking substantial Apache hunting areas only in central Nebraska. On the Southern Plains, they moved only as far as eastern Oklahoma (Hyde 1959:45; Secoy 1953:80).

These three points suggest that factors in addition to the historical pattern of European expansion and trade to the Plains Indians affected the pattern of human movement into the grasslands. The next chapter considers one such factor—the relationship between the complexity of a human society and the nature of that society's environment.

Chapter 8

Ecological Relationships in Recent Plains Society

Because descriptions of the Plains tribes prior to the nineteenth century tend to be extremely sketchy, the relationship proposed here between human organization and environmental conditions is tested most reliably on data from the 1800s. Furthermore, it is reasonable to assume that the data available in the ethnographic record should be most applicable to the more recent periods of the tribes' histories and should be progressively less accurate farther back in time. The focus of this section of the analysis is therefore on the decade from 1840 to 1850, as the most recent period that is likely to represent a reasonably intact aboriginal way of life. The discussion in the preceding chapter indicates that before this time, the tribes had not settled into their final locations; after this time, government handouts, increasingly severe conflicts with whites, and the progressive destruction of the bison herds fundamentally altered the Plains way of life.

The tribes considered here include the nomadic Plains hunting tribes, those groups that "carried on no horticulture, relied on the buffalo as [their] principal means of subsistence, and possessed the horse" (Oliver 1962:13). Following Oliver (1962:19), this definition includes the Arapahoe, the Assiniboine, the three divisions of the Blackfoot (the Blackfoot, the Blood, and the Piegan), the Cheyenne, the Comanche, the Plains Cree, the Crow, the Gros Ventre, the Kiowa and the Kiowa-Apache, the Sarsi, and the seven divisions of the Teton Dakota (the Blackfoot, the Brule, the Hunkpapa, the Miniconjou, the Oglala, the Sans Arc, and the Two Kettles). For convenience, the term *Kiowa* will refer here to both the Kiowa proper and the Kiowa-Apache because these two ethnic groups formed a single political unit: the Kiowa-Apache, for example, had a formal place within the Kiowa tribal circle. This analysis excludes "peripheral" groups such as the Kutenai and the Flathead on the grounds that their subsistence depended heavily on regions other than the Plains and also does not consider the horticultural Plains groups such as the Pawnee or the Mandan.

1. MEASURING RESOURCE AVAILABILITY

The true Plains hunting groups relied primarily on the bison for food and for many other important items, and the ethnographic record makes it clear that they attained their maximum degree of social complexity seasonally, during aggregations linked to the communal hunt. This does not imply that these groups were wholly dependent on the bison—they universally hunted other animals, gathered wild plants, and traded with their horticultural neighbors for corn—but it does imply that these other elements of their diet were not linked directly to the tribal social structure, as were the bison. As the previous discussion has shown. the variation in bison adaptations that is important to this analysis is tied strongly to climatic conditions. Climatic data can therefore be used to construct a rough, indirect measure of the degree to which the habits of the bison in a region encouraged human aggregation and hence favored more heterogeneous societies. More heterogeneous societies should be favored when bison are numerous and tend to aggregate into relatively large, sedentary herds whose movements are both regular within a season and repetitive from year to year. When large aggregations of bison persist for more of the year, humans should also stay together for more of the year, which should also favor greater heterogeneity.

Three climatic variables are particularly important here. The first is total annual precipitation, which controls total forage production and hence ungulate population, herd sizes, and mobility: regional populations and local herds are larger and more sedentary when precipitation is higher than when it is low. Second, the degree of year-to-year variability in annual precipication (measured by the coefficient of variation in mean annual precipitation) determines the similarity of movements from year to year: movements are more repetitive from year to year when variability is low. Last, minimum July temperatures determine the proportion of warm-season grasses in a region, which in turn determines the regularity of grass growth during the growing season and the length of the period during which forage production is high: herd movements are more regular and large herds can support themselves longer in regions with lower July temperatures and thus more cool-season grasses. Heterogeneity should be high when precipitation is relatively high and when annual variation in precipitation and July temperatures are relatively low.

These three variables can be combined into a single index measuring the climatic conditions in any part of the Plains. If P is mean annual precipitation, CV is the coefficient of variation in this precipitation, and J is minimum July temperature, this index is:

$$P/(CV \times J)$$

Increases in this index should favor greater heterogeneity, and decreases in it should favor less heterogeneity.

This index should be considered to be a first approximation to a precise quantitative measure of environmental conditions on the Plains. Its major weakness at present is that it does not distinguish between the effects of the CV, measuring variation in forage conditions between years, and July temperatures, measuring variation in forage conditions with a single year: given a single value for annual precipitation, several different combinations of values of CV and J can produce identical values for the index. Nevertheless, the analysis here indicates that this index is useful in its present form.

The general pattern of variation in this index across the Plains is shown in Figure 8-1, and the data from which the index was computed are listed in Table 8-1. Following the arguments in Chapter 7 that the available data show no convincing evidence that the Little Ice Age climate was significantly different from that of the present, modern climatic data are used here. It should be noted, however, that assertions that the Little Ice Age climate was colder or wetter than that of the present have few implications for the present analysis: If precipitation was 10% higher than that of the present, the precipitation values in Table 8-1 can be multiplied by 1.1 without changing the patterns presented here. Values for mean precipitation and July temperature were obtained from the National Environmental Data Service and refer to the period from 1951 to 1980. The coefficients of variation in precipitation were extrapolated from mapped data presented by Hershfeld (1962) and refer to the period from 1931 to 1960. Data on year-to-year variability in precipitation are unfortunately not available for the more recent period, but the similarity between modern differences in precipitation variability and differences in the degree of variability in tree-ring thickness on the Northern and Southern Plains for the preceding several hundred years, which was noted before, supports this procedure. To check for errors in these last data caused by extrapolations from a map, the values in Table 8-1 were estimated, set aside for several months, and reestimated. No substantial differences were noted between these two estimations, indicating that the data in the table are internally consistent. Figure 8-1 shows a clear decrease in the index from northeast to southwest, with the exception of a peak in the Black Hills region. This pattern suggests that a similar decrease in heterogeneity, except in the region around the Black Hills, should be visible in human societies on the Plains.

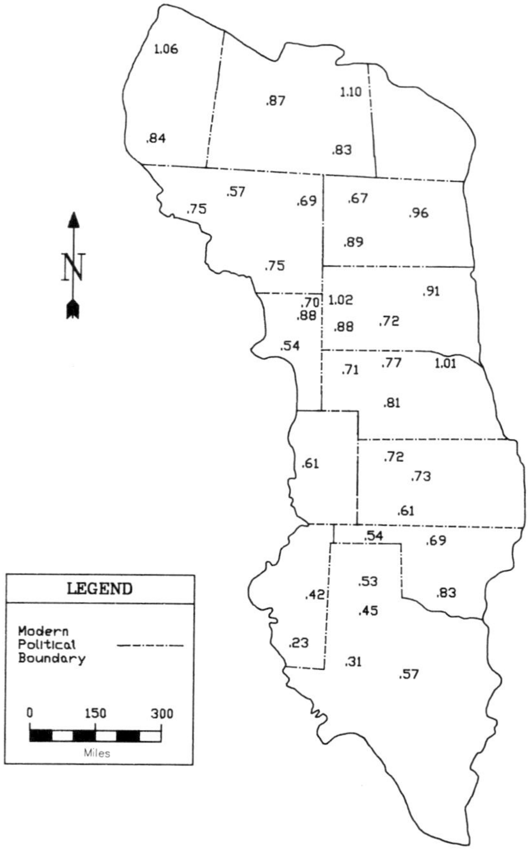

Figure 8-1. Variation in the value of the climatic index across the Great Plains.

2. MEASURING SOCIAL COMPLEXITY

Chapter 2 defined *heterogeneity* as the number of social positions in a society. For this specific analysis, a somewhat more restricted definition is preferable. The process that that discussion proposes rests on the likely inability of a strictly kin-based social structure to mediate disputes and reach and enforce group decisions for large numbers of people. It follows that this analysis emphasizes heterogeneity in the form of non-kin-based tribal social structure that is directly related to formal means of making and enforcing decisions. Among the Plains tribes, there were two principal sources of formal authority: band chiefs and men's, or military, societies.

The Plains tribes were divided into relatively distinct bands, each following a

Table 8-1. Data on Mean Annual Precipitation in Millimeters (*P*), Coefficient of Variation in Mean Annual Precipitation (*CV*), and Mean July Temperature in Degrees Centigrade (*J*) and the Climatic Index for 39 Weather Stations on the Great Plains

Station	*P*	*CV*	*J*	Climatic index
Edmonton, Alberta	450	25	17.0	1.06
Lethbridge, Alberta	400	25	19.0	0.84
Denver, Colorado	394	28	22.9	0.61
Lamar, Colorado	369	33	25.9	0.67
Dodge City, Kansas	525	32	26.7	0.61
Hays, Kansas	568	30	26.1	0.73
Oberlin, Kansas	519	28	25.9	0.72
Syracuse, Kansas	414	31	26.3	0.51
Crow Agency, Kansas	388	26	22.0	0.68
Harlem, Montana	297	25	20.7	0.57
Great Falls, Montana	387	25	20.7	0.75
Opheim, Montana	307	26	18.6	0.65
Imperial, Nebraska	474	27	24.6	0.71
North Platte, Nebraska	495	26	23.4	0.81
O'Neil, Nebraska	580	24	23.9	1.01
Scottsbluff, Nebraska	371	25	23.4	0.63
Valentine, Nebraska	435	24	23.6	0.77
Clovis, New Mexico	419	40	25.1	0.42
Roswell, New Mexico	246	42	27.4	0.23
Dickinson, North Dakota	415	23	20.3	0.89
Minot, North Dakota	454	23	20.5	0.96
Williston, North Dakota	352	25	21.1	0.67
Alva, Oklahoma	632	32	28.6	0.69
Hooker, Oklahoma	489	34	26.8	0.54
Lawton, Oklahoma	742	31	28.7	0.83
Regina, Saskatchewan	380	24	19.0	0.83
Saskatoon, Saskatchewan	360	23	18.0	0.87
Yorkton, Saskatchewan	470	23	18.5	1.10
Custer, South Dakota	461	24	18.1	1.06
Faulkton, South Dakota	457	22	22.8	0.91
Midland, South Dakota	402	23	24.4	0.72
Rapid City, South Dakota	461	23	22.9	0.88
Spearfish, South Dakota	535	24	21.9	1.02
Abilene, Texas	591	36	28.9	0.57
Lubbock, Texas	451	38	26.6	0.45
Midland, Texas	348	41	27.6	0.31
Amarillo, Texas	485	35	26.0	0.53
Casper, Wyoming	290	25	21.6	0.54
Colony, Wyoming	365	23	22.7	0.70
Sundance, Wyoming	430	24	20.4	0.88

specific individual chief. Among many of the tribes, these bands had specific places within the tribal camp circle (e.g., Dorsey 1891), clearly indicating their relatively formal place in the tribe's organization. Although members of any band could, and often did, camp and hunt independently of each other, the band as a whole formed an integrated political unit. Band chiefs held their positions by virtue of their personal qualities, although there was a tendency in many of the tribes for the chieftainship to pass from father to son, and wealthier individuals were more likely to become chiefs than poorer individuals. Chiefs also tended to have fairly limited powers: they usually decided, for example, when and where their band should move camp, and they mediated disputes between individuals and groups, but their decisions were generally based on tribal consensus and were subject to tribal approval (i.e., Flannery 1953:31–36; Grinnell 1962a:338–342).

More coercive power in specific contexts was held by the military societies, which operated seasonally during communal ceremonial and hunting encampments. These groups compelled attendance at tribal ceremonies and policed communal hunts, to ensure that individuals did not drive the bison away by hunting before the tribe as a whole was ready. If they were disobeyed, the members of a society could destroy people's belongings, beat them, or, in extreme cases, kill them. Membership in these societies was independent of band membership, and each society generally had members in all of the bands within a tribe.

Service (1962:12) has identified residential units, including bands of the type found on the Plains, and associations, or sodalities, that crosscut and integrate such units, as the two fundamental components of a group's social structure (the term *sodality* refers here only to nonkin associations, although Service [1962] extends it to include kin-based associations such as clans and moieties as well). One measure of the complexity of such a structure, and hence of the degree of heterogeneity within a society, is the number of positions within it, which for convenience will be referred to here as the number of "social categories."

This number can be derived for the Plains tribes by multiplying together the number of bands and men's societies within each tribe, thereby obtaining the total number of recognized positions within the formal social structure of the tribe into which a man could fit. Women's positions are not considered here because women rarely exercised formal authority on the Plains (Weist 1980). This measure deliberately excludes strictly kin-based social groups, such as the matrilineal clans among the Crow (Lowie 1935); social categories that had no formal role within the tribal authority structure, such as Kiowa social classes (Mayhall 1971:136); and organizations with exclusively ceremonial significance, such as the Dakota dreaming societies (Wissler 1916a:81–91). Other measures of heterogeneity could be constructed, but this one directly measures the complexity of the formal authority structure of the tribes, the social characteristic that is central to this discussion.

It is critical to emphasize here that the values obtained in this way are static, approximate measures of one aspect of inherently dynamic social orders. Not only did bands and societies periodically appear and disappear, but in at least one case, the Cheyenne, the basic tribal social structure seems to have changed over the course of the nineteenth century. Moore (1974) documents a trend among the southern Cheyenne for military societies to withdraw from the traditional tribal structure and form their own bands, apparently as the result of increasingly intense warfare with whites. Although the trend seems to be visible primarily during the 1860s and 1870s, Moore notes a few such withdrawals as early as the 1830s.

As a later section of this chapter notes, the presence of this transformation among the southern but not the northern Cheyenne makes sense not only in terms of the factors that Moore discusses but also in terms of the cultural–ecological relationship proposed here. However, such a change in social structure also emphasizes both the importance of the historical processes operating on the Plains and the ability of the Plains tribes to respond to those changes. The analysis here attempts to take such processes into account by incorporating both nineteenth-century observations and more recent reconstructions of the Plains tribes, but its results must inevitably remain approximate.

The important numerical data for this analysis are therefore the total numbers of politically distinct bands and military or chiefs' societies in each tribe. Chapter 2 also suggests that population pressure is not the major determinant of differences in heterogeneity, and reasonable estimates of tribal population densities are required to test this assertion.

Data are available for all three of these variables, but they must be used with care. Specific bands among the Plains tribes were generally temporary groups following a specific leader. Although the total number of bands existing at any one time may have remained relatively constant, the names, leaders, and members of those bands did not. Lists of band names compiled from informants' memories therefore do not necessarily indicate the number of political units existing at any one time (cf. Ewers 1955:247, for the Piegan). Similarly, military societies formed and disappeared over time, creating similar problems for lists compiled during the early 1900s. Although historical sources and ethnographic work conducted during the 1800s are probably relatively free of this problem, their reliability is often in doubt, particularly when they are based on relatively short trips through the Plains rather than on extended residence there.

Table 8-2 presents data by tribe on the numbers of bands and societies per tribe, noting the source of the information and the date to which it refers. Table 8-3 similarly presents estimates of tribal population. Although there is substantial

Table 8-2. Available Data on Bands and Societies among the Plains Tribes in the Mid-Nineteenth Century

Tribe	Reference	Bands	Societies	Date
Assiniboine	Hodge 1907:104	5, 8, 11	—	—[a]
	Hayden 1862:387	6+	—	1856
	Dorsey 1891	9, 6, 5	—	—
	Lowie 1909a:8, 33–34	17	1	—
	Denig 1930:430–431	6	1	1850
	Denig 1961:79–80	7+	—	1850
	Culbertson 1952:137	2+	—	1839
	Bray and Bray 1976:262–263	9	—	
Arapahoe	Kroeber 1983:5, 154	4	8	—
	Hodge 1907:73	8	—	—
	Trenholm 1970:52–55	4	8	1850s
	Hayden 1862:344	2	—	1856
Blackfoot				
Blackfoot	Culbertson 1952:137	2	—	1850
	Wissler 1912a:21	6	14	—
Blood	Culbertson 1952:137	3	—	1850
	Wissler 1912b:21	7	9	—
	Morgan 1964:151	5	—	1860
Piegan	Culbertson 1952:137	5	—	1850
	Wissler 1912:21	23	9	—
	Morgan 1877:151	8	—	1860
	Ewers 1955:247	7–8	—	—
Cheyenne	Grinnell 1962a:88	10	7	—
	Culbertson 1952:137	3	—	1850
	Hodge 1907:254–256	11	—	—
	Mooney 1907:403, 412	10	6	—
Comanche	Wallace and Hoebel 1952:25–32	13	0	—
	Hodge 1907:328	12	—	—
Cree	Mandelbaum 1940:166–167	14	1	—
	Skinner 1916:517	11–14	—	—
	Ray 1974:185	9	—	1863
	Denig 1961:109–110	10+	—	1855
Crow	Lowie 1935:4, 172–173	3	8	1833
	Culbertson 1952:137	2	—	1850
	Denig 1961:143	3	—	1850
Gros Ventre	Flannery 1953:25, 38–39	12	2	—
	Kroeber 1908:148, 232	8	24	—
Kiowa	Mayhall 1971:134–136	6	6	—
	Hodge 1907:699	6	—	—
	Mooney 1895:227, 229	6	6	1850s
Sarsi	Jenness 1938:10–11	5	5	—
Teton Dakota				
Blackfeet	Culbertson 1952:135	5	—	1850
	Dorsey 1891	6	—	1850
Brule	Culbertson 1952:135	9	—	1850
	Dorsey 1891	13, 18	—	1880–1884

Table 8-2. *(Continued)*

Tribe	Reference	Bands	Societies	Date
Teton Dakota *(continued)*				
Hunkpapa	Culbertson 1952:135	7	—	1850
	Dorsey 1891	9	—	1880
Miniconjou	Culbertson 1952:136	4	—	1850
	Dorsey 1891	9	—	1880
Oglala	Culbertson 1952:136	5	—	1850
	Dorsey 1891	7, 21	—	1879–84
	Walker 1982:21, 32–33	10	7+	—
	Wissler 1916a:7, 13–74	4	10	—
Sans Arc	Culbertson 1952:136	3	—	1850
	Dorsey 1891	7	—	1880
Two Kettles	Culbertson 1952:136	2	—	1850
	Dorsey 1891	2	—	1880

[a]Dashes in this column indicate no specific date given.

agreement on the numbers presented for several of the tribes in these tables, there is substantial disagreement for others. However, several considerations provide a basis for deriving reasonable estimates of band, society, and population numbers for all of the tribes.

First, Dorsey (1891:262–263) notes a large increase in the number of named bands among the Oglala division of the Teton Dakota during the early 1880s, which Hyde (1937:315) states was the result of efforts by the United States government to break the authority of the few strong traditional chiefs. This implies that band totals that refer to periods after the tribes were placed permanently on reservations in the late 1800s and that are substantially larger than earlier totals should be discounted for the period of interest here. Second, travelers passing through the Plains on relatively short visits (such as Culbertson [1952]) are likely to encounter fewer groups and informants than long-term occupants of the region and are therefore likely to record low estimates in their writings. Minor variations in the recorded totals that remain after these two factors have been taken into account may then represent fluctuations in the actual numbers of bands, societies, and people among the various tribes at different times.

Table 8-4 presents the estimates of band and society totals used in this analysis along with the values of the climatic index selected for the various tribal territories. These latter values are discussed later. Band totals generally take the most reliable estimate within the range of reasonable values, following Wissler (1912b:21) for the Blood and Blackfoot divisions of the Blackfoot tribe and Morgan (1964:151) for the Piegan; a rough average of 13 for the Comanche after Wallace and Hoebel (1952:25–32); Lowie (1935) for the Crow; Flannery (1953) for the Gros Ventre; Mayhall (1971), Hodge (1907), and Mooney (1895:229) for

Table 8-3. Available Data on the Population of the Plains Tribes in the Mid-Nineteenth Century

Tribe	Reference	Population	Date
Assiniboine	Hayden 1862:387	2,400	1857
	Lowie 1909a:8	8,000–10,000	1823
	Lowie 1954:13	8,000	1829
	Denig 1930:430–431	4,800	1850
	Denig 1961:109	4,000–5,000	1850
	Schoolcraft 1855:494	5,000	1850
	Schoolcraft 1858:686	8,900	1850s
	Schoolcraft 1858:695	7,000	1852
	Schoolcraft 1858:696	6,860	1847
Arapahoe	Schoolcraft 1851:523	3,500	1847
	Schoolcraft 1855:495	3,500	1850
	Schoolcraft 1858:686	3,000	1850s
	Schoolcraft 1858:695	2,500	1852
Blackfoot			
Blackfoot	Hayden 1862:249	2,400, 2,450	1856
	Ewers 1955:21	2,400	1860
	Lowie 1954:12	2,400	1855
Blood	Hayden 1862:249	1,750	1856
	Ewers 1955:21	1,300	1860
	Lowie 1954:12	2,000	1855
	Schoolcraft 1858:686	1,612	1850s
Piegan	Hayden 1862:249	3,700, 2,520	1856
	Ewers 1955:21	3,700	1860
	Lowie 1954:12	3,200	1855
Cheyenne	Culbertson 1952:137	3,000	1850
	Hayden 1862:276	3,250	1822
	Mooney 1907:402	3,460	1822
	Schoolcraft 1851:523	2,500	1847
	Schoolcraft 1855:494	2,500	1850
	Schoolcraft 1858:686	2,345	1850s
	Schoolcraft 1858:695	2,000	1852
	Schoolcraft 1858:696	2,536	1847
	U.S. Commission on Indian Affairs 1869:394, 462	3,450	1869
Comanche	Wallace and Hoebel 1952:32–32	15,000–20,000	1800–1850
	Wallace and Hoebel 1952:32–32	4,700	1866
	Levy 1961:22	2,000–3,000	1840s
	U.S. Commission on Indian Affairs 1869:384	2,538	1869
Cree	Denig 1961:79–80	3,000–3,500	1855
	Hayden 1862:237–238	4,000–4,500	1856
	Lowie 1954:12	4,000	1835
Crow	Ewers 1955:25	4,500	1833
	Culbertson 1952:137	4,800	1856
	Schoolcraft 1851:523	4,000	1847
	Schoolcraft 1855:494	4,500	1850

Table 8-3. *(Continued)*

Tribe	Reference	Population	Date
Blackfoot *(continued)*			
	Schoolcraft 1858:686	4,650	1850s
	Schoolcraft 1858:695	4,000	1852
	Schoolcraft 1858:696	5,300	1848
Gros Ventre	Hayden 1862:344	2,520	1853
	Ewers 1955:25	2,500	1855
	Schoolcraft 1858:687	2,500	1850s
	Schoolcraft 1858:695	2,500	1852
	U.S. Commission on Indian Affairs 1869:292	2,000	1869
Kiowa	Ewers 1955:25	1,750	1869
	Mooney 1895:227, 229	1,600–1,800	1840s
	Schoolcraft 1851:523	2,000	1847
	Schoolcraft 1858:687	1,918	1850s
	U.S. Commission on Indian Affairs 1869:385	2,216	1869
Sarsi	Jenness 1938:10–11	700	—[a]
Teton Dakota			
Blackfeet	Schoolcraft 1855:494	4,500	1850
	Denig 1961:14	750	1853
	Denig 1961:28	1,100	1833
	Bray and Bray 1976:260	500[b]	1839
	U.S. Commission on Indian Affairs 1869:330–331	900	1869
Brule	Schoolcraft 1855:494	4,000	1850
	Denig 1961:14	2,500	1833
	Denig 1961:28	2,400	1955
	Bray and Bray 1976:261–262	1,500[b]	1839
	U.S. Commission on Indian Affairs 1869:330–331	3,000	1869
Hunkpapa	Schoolcraft 1855:494	3,000	1850
	Denig 1961:28	1,500	1825
	Denig 1961:28	1,400	1853
	Denig 1961:15	750	1833
	Bray and Bray 1976:261	500+[b]	1839
	U.S. Commission on Indian Affairs 1869:330–331	2,000	1869
Miniconjou	Schoolcraft 1855:494	1,000	1850
	Denig 1961:14	1,300	1833
	Denig 1961:25	1,000	1856
	Denig 1961:25	1,125	1853
	Bray and Bray 1976:260	900[b]	1839
	U.S. Commission on Indian Affairs 1869:330–331	2,000	1869
Oglala	Schoolcraft 1855:494	3,200	1850
	Schoolcraft 1858:687	1,500	1850

(continued)

Table 8-3. *(Continued)*

Tribe	Reference	Population	Date
Teton Dakota *(continued)*			
Oglala	Schoolcraft 1858:695	1,500	1852
(continued)	Denig 1961:14	1,500	1833
	Denig 1961:22	1,500	1825
	Denig 1961:22	2,000	1850
	Denig 1961:22	2,500	1856
	Bray and Bray 1976:261	1,500[b]	1839
	U.S. Commission on Indian Affairs 1869:330–331	2,000	1869
Sans Arc	Schoolcraft 1855:494	600	1850
	Denig 1961:15	500	1833
	Denig 1961:28	800	1853
	Denig 1961:28	850	1856
	Bray and Bray 1976:260	550[b]	1839
	U.S. Commission on Indian Affairs 1869:330–331	1,500	1869
Two Kettles	Denig 1961:15	500	1833
	Bray and Bray 1976:260	400[b]	1839
	U.S. Commission on Indian Affairs 1869:330–331	1,500	1869

[a]Dashes in this column indicate no specific date given.
[b]This total is the number of lodges recorded per tribe by Nicollet in 1839 (Bray and Bray 1976) multiplied by Denig's (1961:15) estimate of five Teton Dakota per lodge.

the Kiowa; and Jenness (1938) for the Sarsi. A total of 14 bands is used for the Cree, in light of the agreement between Mandelbaum (1940), Skinner (1916), and Hayden's (1862:237–238) list of 10 major bands plus "several" smaller ones. For the divisions of the Teton Dakota except for the Oglala, the midpoint between Culbertson's (1952:135–136) totals, taken as minimum estimates, and Dorsey's (1891:260–263), taken as maximum estimates, was used, rounded to the nearest whole number. Dorsey's 1884 total of 21 Oglala bands was excluded here as the result of reservation conditions. The number of bands among the Assiniboine is taken as 8, ruling out Lowie's (1909a:8) total of 17 and Culbertson's (1952:137) incomplete total of 2 and taking the midpoint of the range of the remaining numbers.

Although the Cheyenne are generally treated as a single tribe in most descriptive ethnographies (e.g., Grinnell 1962a,b; Hoebel 1978; Mooney 1907), Chapter 7 points out that they divided into northern and southern divisions in the 1820s. These divisions did not meet together between 1838 and 1865, and their separation was sufficiently complete that they developed distinctive differences in dialect and styles of dress during this time (Hyde 1968:336–340). They are therefore taken here as separate groups. There appear to have been 10 Cheyenne bands

Table 8-4. Values for Bands, Societies, Horse Wealth, and Climatic Index Used for Analysis

Tribe	Bands	Societies	Horses/person	Climatic index
Arapahoe	4	9	1.4	0.61
Assiniboine	8	2	0.4	0.67
North Blackfoot	6	14	1.1	1.06
Blood	7	10	1.1	0.75
Piegan	7	9	1.1	0.84
Northern Cheyenne	5	7	1.4	0.63
Southern Cheyenne	5	6	1.4	0.51
Comanche	13	1	2.8	0.57
Crow	3	9	1.9	0.68
Cree	14	2	—[a]	1.10
Gros Ventre	12	3	0.3	0.65
Kiowa	6	6	2.8	0.69
Sarsi	5	5	—	1.05
Brule	11	11	1.1	1.06
Blackfoot Dakota	6	11	0.6	0.87
Hunkpapa	8	11	0.5	0.87
Miniconjou	7	11	0.8	1.02
Oglala	6	11	0.8	0.88
Sans Arc	5	11	0.6	0.87
Two Kettles	2	11	0.6	0.72

[a]Dashes indicate no data available.

before the split (e.g., Grinnell 1962a:88–90), and Dorsey (1905:13) lists 5 Southern Cheyenne bands, implying that the tribe divided approximately in half; Table 8-4 thus takes a total of 5 bands for both divisions. Of the total of 7 societies known for the Cheyenne as a whole, only 5 appear to have been present among the southern bands, and the Wolf Society was not present among the northern bands (Dorsey 1907; Lowie 1916c; Mooney 1907); Table 8-4 therefore tabulates 5 societies (plus a category for nonmembers) among the Southern Cheyenne and 6 (plus nonmembers) among the Northern Cheyenne.

The total numbers of military and chiefs' societies among the other tribes in Tables 8-2 and 8-4 are drawn largely from ethnographic research conducted during the early 1900s. The major exception to this is for the Crow, whose societies were tabulated in 1833 (Lowie 1935:172–173). The societies of the Piegan were also tabulated in 1833 (Wissler 1916b:365), but more recent research identified several societies that must have existed at that time but that were not recorded (Wissler 1916b:366–367, 382, 388). The more recent totals are therefore used here. The Assiniboine had one military society that did not include all men in the tribe (Lowie 1909a:35), thereby creating two distinct categories of members and nonmembers. Although most of the totals in Table 8-4 are relatively well-established, those for the Gros Ventre and the divisions of the Teton Dakota require some discussion.

Three aspects of Gros Ventre social and ceremonial organization must be distinguished here. The first is a series of age-graded sacred dances that are extremely similar to the dances performed by Arapahoe military societies (Lowie 1916c:930–931), presumably because of the historical relationship between the two tribes mentioned in the preceding chapter. However, despite the similarities between the regalia and rules for performing Arapahoe and Gros Ventre dances, the relation of these dances to the political organization of the two tribes was fundamentally different.

Among the Arapahoe, "each . . . dance was correlated with a specific society, and with that society only" (Lowie 1916:933). In contrast, Gros Ventre dances were not specifically linked to any society, and the people performing a dance at any given time formed no formal part of the social structure:

> The two Soldier Societies, the Stars and the Wolves, and their respective constituent companies were permanent organizations. The specific individuals who vowed or joined in any given dance did not constitute such. Their grouping was a purely temporary one, lasting only for the duration of the dance. During, for instance, the Crazy Dance, all who joined in it, both members of the vower's company and others, were called Crazy Dancers, but once the dance was over, they ceased to be such; there was no permanent organization or society of Crazy Dancers. (Cooper 1956:175–176)

Permanent associations of men among the Gros Ventre took two forms. The first was a series of age-graded companies. Kroeber (1908:232) recorded 25 of these and initially considered them to be societies comparable to those found in other Plains tribes. Later work by Flannery (1953:38–39), though, indicates that these 25 companies were not independent societies, but rather were less formal groups of men of similar ages who banded together as boys before entering one of only two formal societies in the tribe—the Stars and the Wolves. These two societies formed the second kind of permanent social group within the tribe.

The companies and the societies seem to have played very different roles in the tribal authority structure. Members of an age-company could vow the various dances in the Gros Ventre sequence, and the company as a whole would then put the dance on, but the major tribal police duties appear to have been in the hands of the two larger societies. Flannery (1953:44), for example, states that the rules to be followed during a dance were specified and enforced by either the Wolves or the Stars as a whole, and Cooper (1956:180) states further that if any of the Gros Ventre failed to appear for a dance that required the presence of the whole tribe, "the vower would request the soldiers of the Stars or the Wolves, as the case might be, to compel attendance." This view agrees with Stewart (1977:325), who also suggests that police duties among the Gros Ventre were not linked to the age-companies. The total number of societies in Table 8-4 reflects this interpretation.

In opposition to this view, Fowler (1982:83) emphasizes the similarities between the dances of the Arapahoe military societies and the sacred dances of the Gros Ventre. Relying on these similarities, Fowler argues that the sequence of Gros Ventre dances corresponded to a formal series of ceremonial societies that was crosscut by the Star and Wolf societies and states that "the age-group systems of the Arapahoes and the Gros Ventres served as an overarching political structure during the spring and summer when the sacred lodges and communal hunts were held." The preceding discussion should show that this view does not seem to fit the available ethnographic data.

The total of 10 societies for the divisions of the Teton Dakota represents the 6 *akicita* or police societies, 2 civil societies, and 2 warrior societies recorded for the Oglala (Wissler 1916a). The Ska Yuha chief's society (Wissler 1916a:41) is excluded because it was formed after the Oglala were placed on a reservation. The *blotaunka* (Wissler 1916a:54–61) are also excluded because this term refers not to a formal, permanent association of men but rather to a position on large war parties that was temporarily filled by different men and that existed only for the duration of a particular expedition. This total of 10 is extended to the other 6 divisions of the Dakota because there are no comparable data for them. Culbertson's (1952:78) statement in 1850 that the men's societies were represented in "all parts of the Sioux nation" provides some support for this decision.

An additional category is added to the total number of societies for the Arapahoe, Assiniboine, Cheyenne, Cree, Crow, Gros Ventre, and Teton Dakota, to account for the men of these tribes who were excluded from society membership (Denig 1930:436; Flannery 1953:38; Grinnell 1962b:48; Lowie 1916b:842; Mandelbaum 1940:224–225; Trenholm 1970:77; Wissler 1916a:64). Among the Kiowa and the Sarsi, all men were members of some society, the Comanche had no societies, and the available sources on the Blackfoot do not state clearly whether all men were members or not. The total number of male social categories in a tribe is then obtained by multiplying together the number of bands and the number of societies.

Each tribe in Table 8-4 is also represented by a value of the climatic index discussed before, which is computed for a specific point within its territory. Although the Plains tribes were extremely mobile and boundaries between allied tribes were often blurred (i.e., Sharrock 1974), contemporary observers were uniformly able to identify regions that were recognized not only by whites but also by natives as the habitual ranges of identifiable ethnic groups. The value of the index used for the tribes other than the Teton Dakota and the Crow is the highest available in the approximate center of the tribal territory. The point in southeastern Montana taken for the Crow follows Frison (1967, 1978:234) who identifies this region as the major area for Crow communal hunts. The highest values of the index in the center of Dakota territory are in the Black Hills, but this region is not at the center of any of the territories of the seven tribal divisions. However, all of the divisions except for the Two Kettles appear to have included portions of the Black Hills in their habitual ranges.

Although the Two Kettles appear to have stayed in the Plains between the Black Hills and the Missouri River, there is some disagreement among contemporary sources about the locations of the specific territories of several of the other six Dakota divisions during the 1840s. Most authors place the Oglala in the southwest of Dakota territory, the Brule in the southeast, and the Miniconjou on the northeast (i.e., Bray and Bray 1976:260–261; Culbertson 1952:135–136; Hayden 1862:372–374), the exception being Warren (1856) who locates the Miniconjou on the northwestern edge of the Black Hills. In 1839, Nicollet placed the Sans Arc northeast of the Black Hills, with the Blackfoot and the Hunkpapa north and west of them (Bray and Bray 1976:260–262).

Similarly, in 1850 Culbertson (1952:135–136) found the Hunkpapa and the Blackfoot divisions together in the north, as did Warren (1856). Culbertson, though, places the Sans Arc in the same region as the Miniconjou, whereas Warren locates these two groups ajacent to one another with the Sans Arc directly in the Black Hills and the Minconjou just to the north. Also in 1850, Schoolcraft (1855:494) placed the Miniconjou, the Hunkpapa, the Oglala, and the Sans Arc in unspecified portions of the Black Hills, with the Brule just to the south and the Blackfoot to the northeast. In contrast, Hayden (1862:372–374) does not mention the location of the Blackfoot or the Sans Arc but places the Hunkpapa together with the Miniconjou.

Taken together, these data suggest that all of the Dakota divisions inhabited the Black Hills region regularly, with the apparent exception of the Two Kettles. Hassrick (1964:74), in fact, notes that the Dakota referred to this region as their "Meat Pack." The locations of the Oglala on the southwest, the Brule on the southeast, and the Miniconjou on the northeast are fairly clear. The lack of concensus about the habitual ranges of the Hunkpapa, the Sans Arc, and the Blackfoot suggests that these divisions may have been less strongly tied to specific areas. Hyde (1937:38–39) states that the Hunkpapa and the Blackfoot were among the last of the Dakota groups to move west of the Missouri River, and they may therefore not have been in the Black Hills region long enough to settle in one more or less specific region.

For all of the Dakota groups except the Two Kettles, the value for the climatic index used here is taken for a station on the edges of the Black Hills; the interior of this region is not used because the bison did not range in any great numbers into the higher portions of the Hills themselves (Denig 1961:6). Values for single stations are used for the Oglala, Brule, Miniconjou, and Two Kettles. For the Hunkpapa, Sans Arc, and Blackfeet, who apparently ranged throughout most of the northern part of Dakota territory, an "average" value is used, obtained by taking the mean of the values computed for Sundance (0.88) and Colony (0.70), Wyoming, and Spearfish, South Dakota (1.02).

Table 8-5 presents the values used for tribal populations and areas, along with the population density estimates derived from them. Assiniboine population was substantially reduced by smallpox in 1838 (Hayden 1862:381), making earlier estimates too high. Excluding these estimates, a value for Assiniboine population was derived by taking the midpoint of the range of estimates. Population figures for other groups are generally not too divergent, and the values used here are the midpoints of the range of available estimates, as for the Assiniboine. For the two Cheyenne divisions, this value was simply divided in half. The midpoint used for the Hunkpapa excludes Nicollet's explicitly incomplete estimate of 500 people (Bray and Bray 1976:261). The Comanche are the only group for which this procedure is likely to produce substantially biased estimates. Maximum estimates of Comanche population in the first half of the 1800s range as high as 20,000, with none less than 12,000. Levy (1961) points out that it is extremely difficult to reconcile these estimates either with environmental conditions on the Southern High Plains or with the actual numbers of Comanches who finally enrolled on the reservations in the 1870s.

Table 8-5. Values for Population, Tribal Territory, and Population Density Used for Analysis[a]

Tribe	Population	Territory	Density
Arapahoe	3,000	40,000	0.08
Assiniboine	5,650	140,000	0.04
North Blackfoot	2,425	70,000	0.04
Blood	3,110	70,000	0.04
Piegan	1,525	60,000	0.03
Northern Cheyenne	1,375	22,000	0.06
Southern Cheyenne	1,375	35,000	0.04
Comanche	3,500	150,000	0.02
Crow	4,650	80,000	0.06
Cree	4,650	170,000	0.03
Gros Ventre	2,260	67,000	0.03
Kiowa	1,908	46,000	0.04
Sarsi	700	40,000	0.02
Brule	2,750	30,000	0.09
Blackfoot Dakota	2,350	—[b]	0.10
Hunkpapa	1,875	—[b]	0.10
Miniconjou	1,450	26,000	0.05
Oglala	2,350	26,000	0.09
Sans Arc	1,000	—[b]	0.10
Two Kettles	950	13,000	0.07

[a]Tribal territory in square kilometers; population density is population/square kilometer.

[b]Total Blackfoot Dakota/Hunkpapa/Sans Arc territory is 50,000 square kilometers.

Authors who present these high estimates offer no explanation for the apparent decrease of as much as 80% in Comanche population during the mid-nineteenth century with no parallel decrease in the intensity of Comanche raids on neighboring white settlements. In addition, contemporary observers discussing Comanche population as late as the 1850s explicitly place little confidence in their estimates (i.e., Schoolcraft 1852:125; 1853:635). These extreme estimates are therefore excluded here, and the midpoint taken for the analysis is that between a government population estimate from 1866 and Levy's (1961) minimum of 2,000. In addition, Schoolcraft's (1855:494) estimate of 4,500 Blackfoot Dakota is substantially out of line with other population estimates for this group. It is therefore unlikely to be correct and is excluded from this analysis.

The area estimates used to compute population density figures are derived from Figure 7-3, except for those for the divisions of the Blackfoot and the Teton Dakota, which were derived from Figures 8-2 and 8-3. These data can be taken

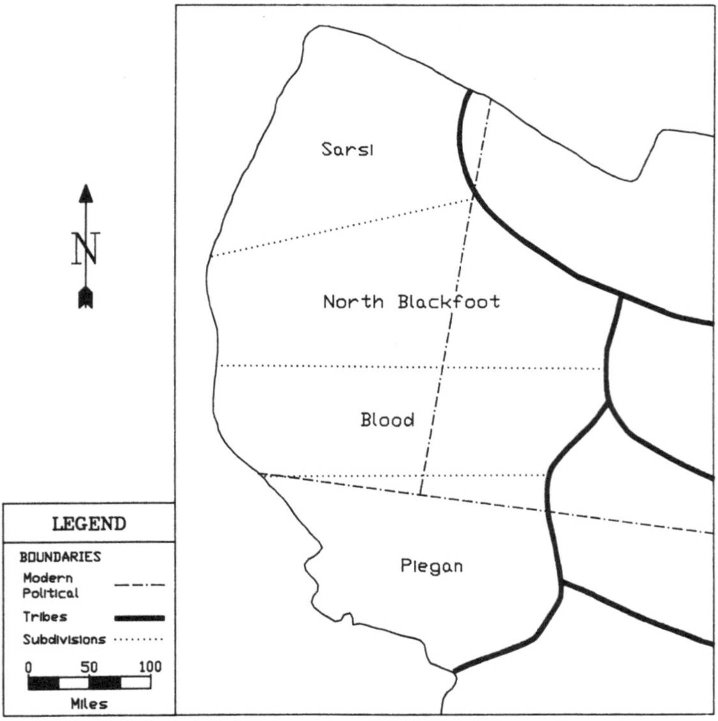

Figure 8-2. Locations of tribal subdivisions within Blackfoot territory.

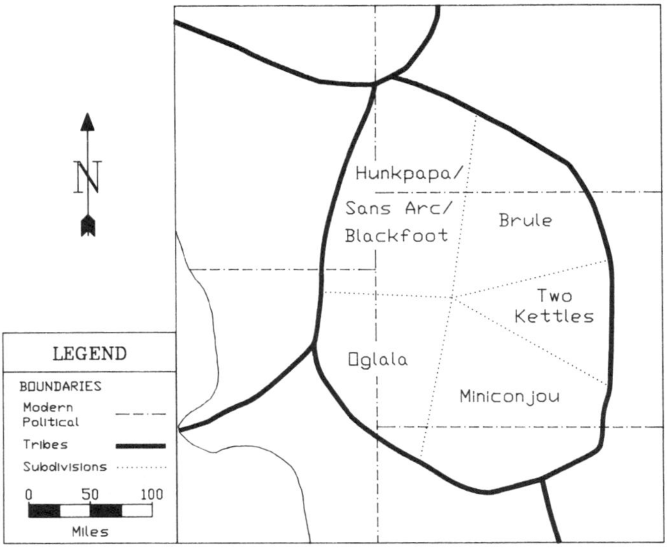

Figure 8-3. Locations of tribal subdivisions within Dakota territory.

only as rough approximations of the size of the area regularly used by each of the different tribes: The boundaries presented here do not mark precise territorial limits, and the great mobility of all of the Plains tribes allowed them to hunt and camp with distant allies (cf. Levy 1961:21). Estimates of population densities computed from these data are therefore similarly approximate. Density figures for the Hunkpapa, Blackfoot Dakota, and Sans Arc, who apparently occupied a single area, were obtained by summing their populations and dividing by the total areas indicated in Figures 7-3 and 8-3.

3. ANALYSIS

Figure 8-4 plots the values of the climatic index and the total number of male social categories, or the degree of social heterogeneity, for each tribe. This plot conforms to the prediction of the major hypothesis presented in Chapter 2—that heterogeneity should increase as environmental conditions increasingly favor human aggregation. Excluding two obvious outliers, the Cree and the Sarsi, the correlation between the climatic index and the number of categories among the Plains tribes is 0.88 ($df = 16$, $p > .001$). The two tribes that clearly depart from this pattern provide some additional insights into the relationship at issue here.

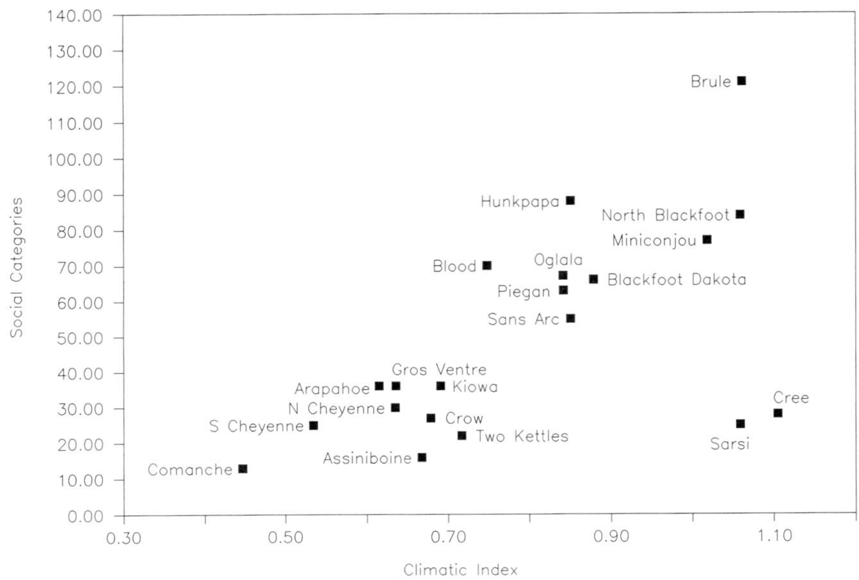

Figure 8-4. Plot of social categories against the climatic index.

Both of these groups show unusually low heterogeneity for the environments they inhabited, probably for two distinct reasons. First, the Sarsi were the smallest of the tribes considered here (Table 8-5). Regardless of the degree to which an environment favors heterogeneity, population size must set a limit on the number of social categories that are needed or can be supported, and the Sarsi population may not have been large enough to support a more heterogeneous social structure. This can be taken into account, and population can essentially be held constant, by dividing the number of social categories present in a given tribe into the total population for that tribe, to obtain a relative measure of social complexity, which can be referred to as "population per category": relatively more heterogeneous societies should have more categories with fewer people in them than relatively less heterogeneous societies. When this analysis is done (see later text), the Sarsi do not appear to be an exception to the general pattern indicated by the other tribes, lending some support to this possibility. It is also possible that the low heterogeneity of this tribe reflects the very simple organization of their Athabascan relatives (cf. Jenness 1977:118–126).

The Cree present a very different situation, illustrating the role of technology in cultural–ecological relationships. The Cree were the poorest tribe in horses on the Plains (Ewers 1955:22–28; Mandelbaum 1940:194–195), a condition that greatly hindered their ability to procure bison. Lacking horses, the effective density of the bison was lower for the Cree than it would have been for better mounted groups, and the effective mobility of the herds was higher. The potential of a given environment to support a given human group can only be assessed relative to the ability of that group to exploit that environment, and, in this sense, the Plains Cree are not truly comparable to the better mounted Plains tribes. In addition, the Cree remained relatively active in the fur trade well into the nineteenth century (Ray 1974), an activity that tends to disperse human populations and maintain a simpler, more flexible social organization. This may also have reduced Cree heterogeneity as it is measured here.

The importance of the interaction between the three climatic variables incorporated into the index used here is illustrated in Figures 8-5 through 8-7, which plot male social categories against mean annual precipitation, the coefficient of variation in mean annual precipitation, and mean July temperatures, respectively. None of these figures shows a strong relationship, although the maximum number of social categories found in regions with very variable precipitation and high July temperatures is lower than in more constant and cooler regions. Only when all three variables are combined into the overall index is a strong link between environmental conditions and organizational complexity apparent.

The absence of any relationship in Figure 8-5 is particularly interesting in light of Reher and Frison's (1980:43–50) discussion of changes in organizational

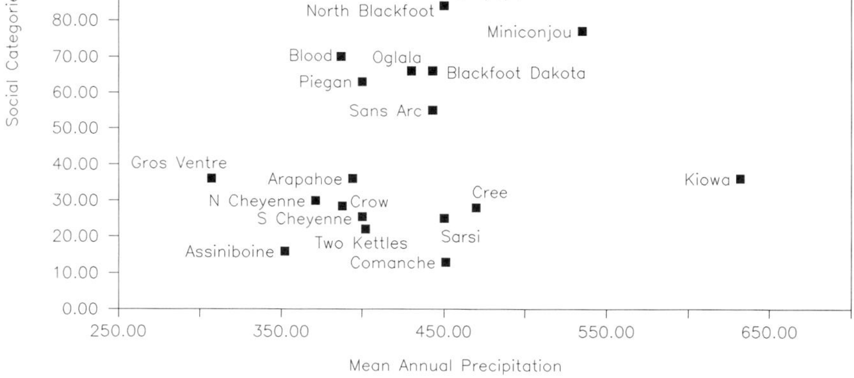

Figure 8-5. Plot of precipitation against social categories.

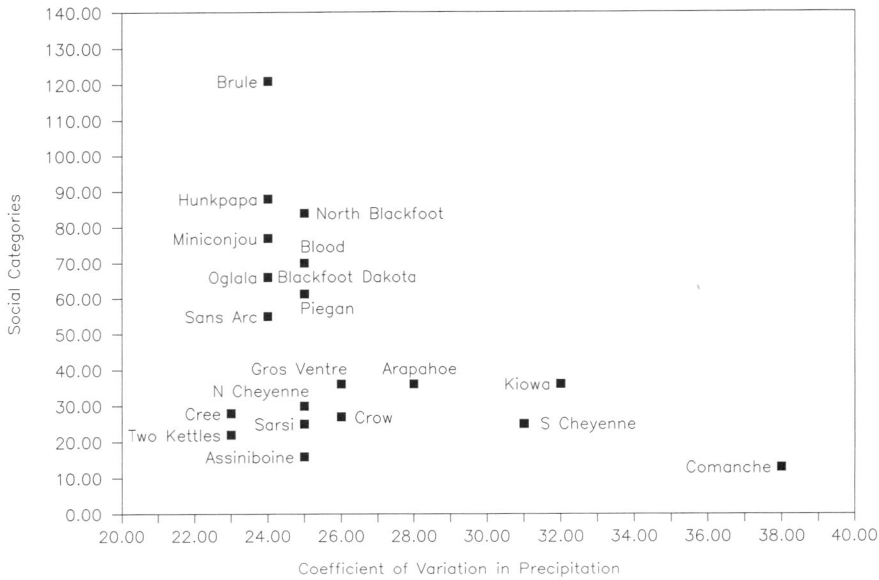

Figure 8-6. Plot of *CV* against social categories.

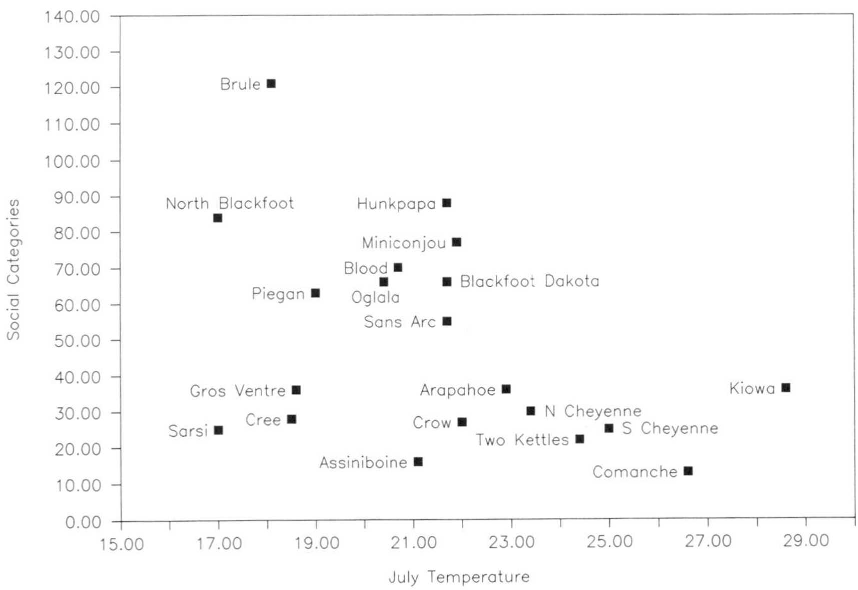

Figure 8-7. Plot of social categories against mean July temperature.

complexity among the Protohistoric inhabitants of the Northwestern Plains. These authors argue that increased precipitation during the Little Ice Age led to higher forage production and therefore larger bison populations. Given these larger populations of prey and the concurrent introduction of the horse, Reher and Frison then argue that human aggregations for communal hunts should have occurred more often and should have been larger, and human organization should have become more complex.

The evidence cited in Chapter 6 indicating that Little Ice Age precipitation was barely greater than that at present causes problems for this account, but in principle the process Reher and Frison propose should have operated on spatial as well as temporal differences in rainfall: human groups in areas receiving more rain and therefore having greater forage production and larger bison populations should have been more complex than human groups elsewhere. Figure 8-5 indicates that they were not. Whether the Little Ice Age was substantially wetter than other periods or not, the data presented here indicate that average annual environmental conditions do not determine human adaptations and that the nature of the seasonal and year-to-year variation in those conditions is critical to cultural–ecological relationships.

Chapter 2 pointed out that social complexity is not necessarily linked to population pressure. and the data available here support this assertion. First,

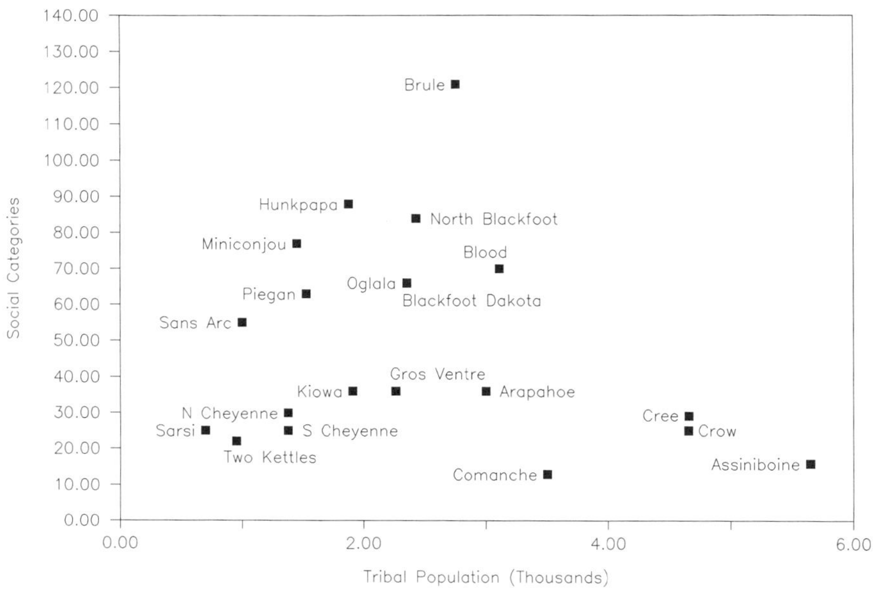

Figure 8-8. Plot of social categories against tribal population.

Figure 8-8 plots total tribal population against the number of social categories for the Plains tribes, and the lack of any linear relationship is apparent ($r = -.20$, $df = 18$, $p > .1$). The ability of larger populations to support more heterogeneous social structures, noted earlier, is evident in the steady increase in the maximum number of social categories associated with larger populations, but population size clearly does not determine heterogeneity. Nor is there a relationship between the climatic index and total population (Figure 8-9; $r = -.17$, $df = 18$, $p > .1$).

Figure 8-10 plots population density against heterogeneity. again showing no relationship ($r = .36$, $df = 18$, $.1 > p > .05$). This pattern may be deceptive, however, because the productivity of the various tribal territories varies substantially, and a given density of human beings does not indicate the same relationship between population and resources in a highly productive region that it does in a minimally productive region. Chapters 2 and 3 showed that faunal productivity on the Plains is a function of precipitation, and one way to take account of regional differences in production is therefore to divide the values for tribal population density by mean annual precipitation. Figure 8-11 plots these adjusted density figures against heterogeneity, showing an even weaker relationship than that in Figure 8-10 ($r = .19$, $df = 18$, $p > .1$). Population pressure therefore appears to be unrelated to heterogeneity as these variables are measured here.

The relationship noted before between population per category and the

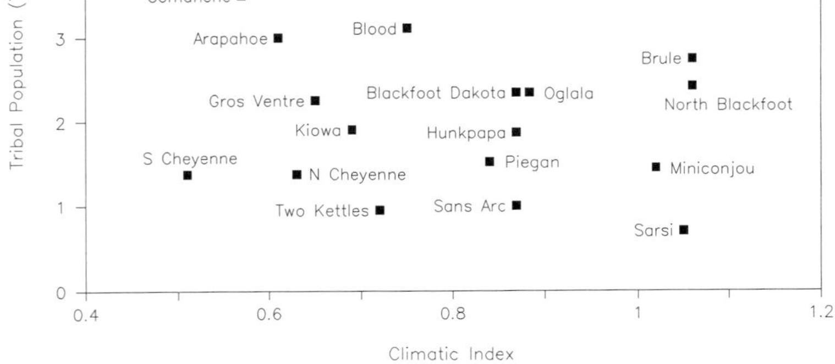

Figure 8-9. Plot of tribal population against the climatic index.

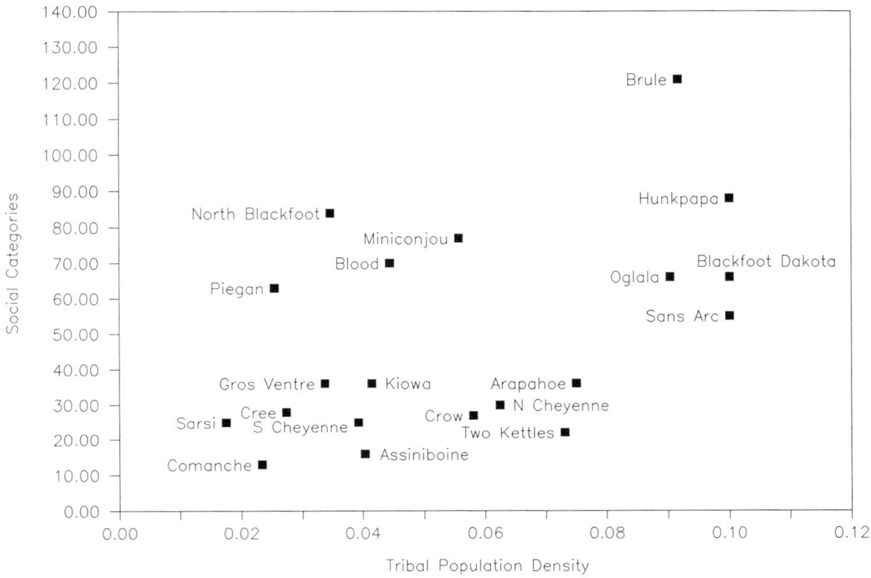

Figure 8-10. Plot of social categories against population density.

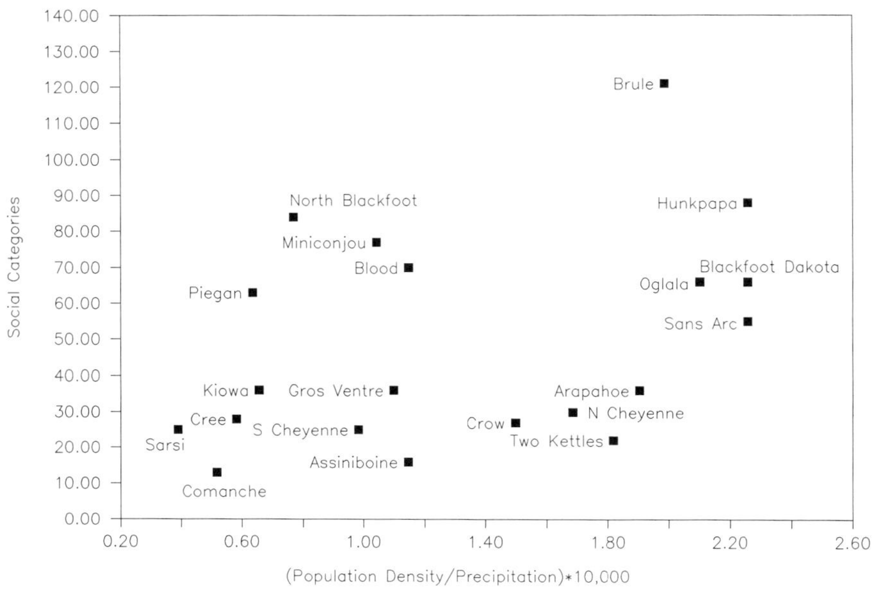

Figure 8-11. Plot of social categories against population density/precipitation.

climatic index, though, is very interesting. Chapter 2 relates social complexity to patterns of human interaction rather than to overall population size or density, and Figures 8-8 through 8-11 show that absolute population size is not an important determinant of heterogeneity, although they suggest that absolute population size does limit the maximum degree of complexity that a society can support. Total population per social category is a measure of complexity that is independent of absolute population size and that should therefore indicate whether societies with different population sizes have attained similar degrees of complexity. Figure 8-12 plots this variable against the climatic index.

This figure demonstrates an obviously nonlinear drop in the number of people per social category as the value of the climatic index increases, with the major exception of the Cree. The position of the Assiniboine suggests that they may also be an outlier on this curve. Like the Cree, the Assiniboine were relatively poor in horses and may therefore not be quite comparable to the other tribes for the reasons discussed before. As was noted earlier, the position of the Sarsi on this plot conforms exactly to the relationship indicated by the other tribes. This relationship can be made linear by taking the natural logarithm of population per category and the inverse of the climatic index. Excluding the Cree, the correlation between the two variables transformed in this way is .79 ($df = 16$, $p < .001$); excluding the Cree and the Assiniboine, the correlation is .83 ($df = 15$, $p < .001$).

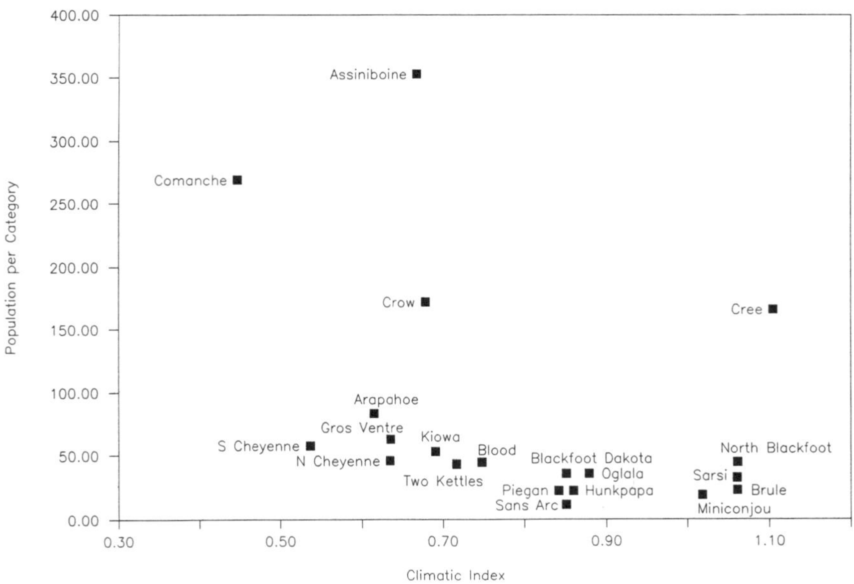

Figure 8-12. Plot of population per category against the climatic index.

It is also possible to rule out an alternative explanation for the observed variation in social heterogeneity among these tribes. In at least some of the Historic Plains tribes, the acquisition of horses in the eighteenth century noticeably increased social inequality, with an individual's wealth being measured in horses (Ewers 1955:28–31). Inequality of this kind is one component of the general concept of social complexity (see Chapter 2 and McGuire 1983), and it might be argued that access to horses was a major determinant of heterogeneity, the component of complexity that is important here. However, this does not appear to be the case.

Ewers (1955:28) provides estimates of horse wealth (horses per person) for most of the tribes considered here (summarized in Table 8-4). Although Ewers's figures are for 1874, several decades later than the period considered here, they are the only reasonably accurate data that are available. Figure 8-13 plots the number of male social categories against the number of horses per person for the 17 tribes for which both values are available. There is clearly no evidence in these data that greater horse wealth increases social heterogeneity ($r = -.26$, $df = 16$, $p > .1$); if anything, the tribes with the most horses per person appear to be simpler than less wealthy tribes.

Finally, the analysis to this point has been strictly bivariate, and it is possible that some variables might affect social heterogeneity only in combination with

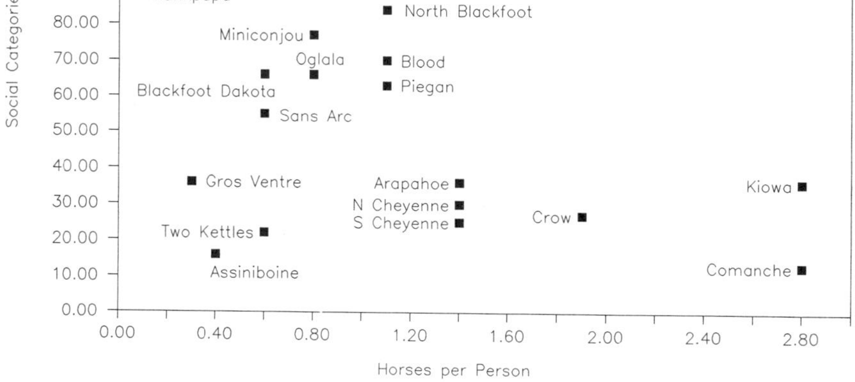

Figure 8-13. Plot of social categories against horse wealth.

Table 8-6. Multiple Regression of Heterogeneity as a Function of (1) the Climatic Index, Raw Population Density, and Horse Wealth, and (2) the Climatic Index, Adjusted Population Density, and Horse Wealth

Variable	Standardized coefficient	*t*	Probability
1. Constant	0.000	−2.415	.031
Climatic index	0.862	5.777	.000
Raw density	0.076	0.510	.619
Horse wealth	0.012	0.077	.940
	Adjusted multiple r^2 = .739		
2. Constant	0.000	−3.114	.008
Climatic index	0.892	6.438	.000
Adjusted density	0.124	0.859	.405
Horse wealth	0.116	0.755	.463
	Adjusted multiple r^2 = .736		

other variables. Table 8-6 shows the result of a multiple regression incorporating the major potentially explanatory variables considered here: the climatic index, raw and adjusted population densities, and horse wealth (this analysis automatically excludes the Sarsi and Cree, outliers on Figure 8-4, for whom Ewers provides no quantitative estimates of horse wealth). The results of this analysis indicate a strong relationship between the climatic index and heterogeneity but show no relation at all involving either measure of population density or horse wealth, confirming the bivariate results mentioned before.

These results are thus consistent with the predictions of the theory presented in Chapter 2: more complex societies are found in regions in which bison herds should have been larger, more sedentary, and more regular in their migrations and that therefore should have favored larger, longer, and more frequent human aggregations. The major weakness in this analysis is simply that it is not possible to observe directly the process proposed here because of the lack of data on the length, size, or regularity of human aggregations held in different parts of the Plains. Comparing tree-ring widths with records of Kiowa tribal aggregations and Sun Dances, Levy (1961) has shown that the tribe remained dispersed in extremely dry years, as the present discussion predicts. However, there are currently no systematic comparative data on the regularity of tribal aggregations elsewhere on the Plains. The pattern shown in Figure 8-4 is the one expected as aggregations became larger, longer lasting, and more regular, though, supporting the discussion in Chapter 2.

To summarize, both the increase in social categories and decrease in population per category as the climatic index increases conform to expectations about the increasing importance of a more complex social structure when more people are in face-to-face contact with one another more often or over longer periods of time. The absence of any relationship between total population and heterogeneity also indicates that it is not simply large numbers of people within a region that demand more complex organizations. Rather, the important factor appears to be the way in which those people are distributed across the landscape. As Chapter 2 notes, Schalk's (1979) data on Northwest Coast society show this same pattern.

4. ECOLOGY AND HISTORY IN RECENT PLAINS SOCIETY

The analyses just presented thus indicate that the organization of human groups on the Plains during the first half of the nineteenth century was strongly linked to the nature of the territories those groups habitually occupied. Although white contact certainly altered the organization of the Plains tribes (e.g., Lewis 1942:40–43; Moore 1974), it seems almost certain that the *relative* differences in complexity among the various tribes reflect pre-Contact conditions. Lewis (1942:41) suggests that white contact increased the number of men's societies among the Blackfoot but that several such societies existed aboriginally. In contrast, neither the Shoshone (Lowie 1909b:206, 216) nor the Comanche (Wallace and Hoebel 1952:33) shows any evidence of having had men's societies, and the "extreme simplicity" (Lowie 1909b:206) of the organization of these groups clearly reflects their Great Basin origins (compare Steward 1938). The continuities that Oliver (1962) demonstrates between the organization of the Plains tribes and the organization of the tribes in the regions from which they migrated to the Plains also indicate that the relative differences in complexity shown here are not a recent development.

This being the case, the relationship just documented between complexity and environmental conditions helps to explain the population movements summarized in Chapter 7. The general pattern of migrations is relatively simple. Prior to the introduction of the horse, the Western Plains from the Canadian–United States border into northwest Texas were dominated by the Shoshone and the Apache. Following the acquisition of the horse by the Shoshone in about 1700, this group expanded to the south at the expense of the Apache. This initial expansion ended on the Arkansas River in approximately 1725.

By 1750, horses and guns had diffused to most or all of the other tribes on the Plains. The northern Shoshone groups were well-supplied with horses but were unable to obtain many guns. However, the southern Shoshone groups, now known as the Comanche, had access to both horses and guns and were well-supplied with both. Between 1750 and 1800, the Shoshone were pushed back into the Rocky Mountains, and the Comanche moved or were pushed south into Texas, completely displacing the resident Apache. These movements were initially caused by the southwestward movements of the Blackfoot, the Crow, the Cheyenne, and the Kiowa, but the appearance of the Teton Dakota west of the Missouri River soon put pressure on these groups as well.

After the introduction of the horse, the overall trend in this sequence is the constant expansion by military conquest of more complex groups at the expense of less complex groups into areas of the Plains that are less suitable for communal hunting and hence for the maintenance of social complexity. The possible exception to this is in the Comanche conquest of the Apache, which was accomplished by massing mounted warriors at small Apache farming villages during the seasons when the Apache had to disperse into the villages to tend their crops. The historical introduction of the horse and the gun set this sequence of movements in motion, but there are several factors that indicate that the cultural–ecological relationship discussed here had an important effect on them.

Osborn (1983) has pointed out that a major effect of the introduction of the horse was an increase in the effective density of the bison in a region and in the ability of hunters to find and kill those bison: a prey species is only "unpredictable" or "mobile" relative to a predator's ability to locate or keep up with it. This means that areas on the Plains that were not suited for more complex pedestrian societies were transformed into areas that were well-suited for more complex mounted societies once horses became widespread.

This transformation was of great importance on the Historic Period western Plains. Although an emphasis on mounted warfare led to an increase in small war parties hoping to obtain personal glory and to steal horses (Lewis 1942:51–58; Secoy 1953:62–64), major conquest and territorial defense depended on numerical strength. The Shoshone, for example, conciously attempted to stay in the largest camps possible after their enemies obtained firearms and made war by trying to attack small enemy camps with overwhelming force. Despite their access to guns, the Blackfoot similarly assembled in large camps for defense (Secoy 1953:52–53; Sutton 1986).

The Teton Dakota in particular had an advantage because their alliance of a large number of distinct groups provided a large pool of warriors for mass attacks. Secoy (1953:76) states that toward the end of the eighteenth century, "horse-raiding was carried on by small, rather informal war parties, while effective domination of a particular territory was established and maintained by formal, large-scale actions." Among the Dakota, these actions involved large numbers of warriors drawn from a single massed encampment; in many cases, old men, women, and children accompanied the warriors in the search for the enemy and were left behind only when a target was located.

Maintaining these large aggregations and integrating large numbers of warriors on a regular basis clearly demands the kinds of mechanisms for social control that are at issue here. Over the long term, societies lacking such mechanisms must therefore have been at a military disadvantage when they were in conflict with societies that had them. Before the introduction of the horse, the Shoshone in particular were able to make a living from areas of the Northwestern Plains in which their enemies probably could not support themselves. Following this introduction, they were unable to maintain their military superiority. The historical accident of the distribution of European trade goods, particularly firearms, was obviously critical to the aboriginal balance of power on the Plains, but this accident occurred in a predictable ecological context that influenced its effects.

This relationship also suggests an explanation for the uniformly northeast-to-southwest pattern of recent tribal movements and the absence of any substantial expansion from the east onto the Southern Plains. The more complex groups who moved onto the Plains, some of whom abandoned agriculture to do so, were those groups adjacent to the areas in which more complex groups could support themselves, and the most complex of these groups on the Plains at any given times seem universally to have lived in the vicinity of the Black Hills, the area most suitable for them. Groups living on the margins of the less productive, less predictable southern parts of the Plains did not expand far into the mixed- or short-grass steppe, probably in part because they could not make a living there and still maintain their social order.

Moore's (1974; also see Hoebel 1980; Moore 1981) data on nineteenth century changes in southern Cheyenne social structure are also interesting in this light. Moore shows that competition between civil and society chiefs probably led increasing numbers of southern Cheyenne military societies to withdraw from the traditional tribal organization and form independent bands of their own, a process that he links to warfare with whites. He also suggests that less intense warfare in the north until much later, along with the ability of some society chiefs to join the northern tribal council without renouncing their society positions, inhibited this process among the other major division of the tribe.

However, these patterns may also have an ecological basis. The southern Cheyenne occupied the region just north of the Comanche, a section of the Plains that is almost as arid and only slightly less variable than Comanche territory. The trend in southern Cheyenne social structure is not only from the traditional matrilocal band system to a new patrilocal, society-based system, it is also to a system without sodalities cutting across band lines, a pattern found in the other Plains tribes only among the Comanche. Particularly before 1860, when raiding began to supplant bison hunting as the major economic support for the tribe (cf. Moore 1974:342), the southern Cheyenne trend may thus reflect an adaptation to the exigencies of making a living in one of the more erratic and unproductive portions of the Plains. Similarly, the very ability of the society chiefs to enter the tribal council in the north may reflect the adaptive advantages of a more complex social structure under the less variable conditions in that region.

On the Historic Period Great Plains, military advantages were obviously very important: warfare was constant, and the ability of any group to hold a place in the region depended largely on its ability to defend itself. However, warfare was probably not as intense throughout prehistory as it was in recent times, and the military benefits of greater social complexity may therefore have been less important in the more distant past. The relationship shown here, though, is not linked only or necessarily to military considerations and should therefore hold whether or not armed conflict is intense. The military advantages of increased complexity depend in part on the ability of large aggregations to support themselves in the Plains environment, which in turn gives the participants in such aggregations an advantage in battle. Several factors contribute to this ability.

The communal hunt itself can be enhanced by larger aggregations of people. Such aggregations bring together information about herd sizes and numbers over a larger region than do smaller aggregations, increasing the chances of locating a herd. They also ensure adequate manpower for the hunt and may offer the possibility of conducting several independent hunts. Particularly in prehorse times, for example, a large group might have been able to work at several pounds or jumps simultaneously. Other considerations are also important. Larger groups present a wider choice of potential marriage partners as well as access to more trade goods obtained from a large area. Although the relationship shown here appears to have had clear effects on success in the armed conflict that was prevalent on the Plains in the eighteenth and nineteenth centuries, the nonmilitary advantages just noted suggest that it should hold even if warfare is not common. There is thus good reason to expect heterogeneity to change over time with resource structure in an evolutionary sequence within a single region.

The ethnographic literature suggests a mechanism by which such a change might occur if bison herding and migration patterns changed over time. The number of people following any particular leader on the Plains depended largely on the ability of that leader to ensure success in war and the hunt: successful leaders had many followers, and unsuccessful leaders had few (Ewers 1955:245–248; Mandelbaum 1940:221; Walker 1982). Fawcett (1987) considers this mechanism in detail, arguing that ambitious individuals within social groups could manipulate the communal hunt to obtain power or prestige, and his analysis of the social process on which this mechanism relies thus provides an important complement to the present emphasis on the ecological context within which this mechanism operates.

If regional environmental conditions became increasingly better suited to successful communal hunting over the long term, following the most successful hunters would tend to bring people together into larger and more permanent groups, which should then lead to greater heterogeneity. If conditions became increasingly more poorly suited to successful communal hunting over the long term, a leader's success in the hunt would decline, and such groups should break up, decreasing heterogeneity. The following chapters therefore consider an example of long-term environmental and cultural change in a single region of the Great Plains, the Southern High Plains of western Texas and eastern New Mexico, during the Paleoindian Period, from approximately 12,000 to 8000 B.P.

Chapter 9

Recent and Paleoindian Environments of the Southern High Plains

To this point, this study has presented a hypothesis about the relationship between variation in ungulate adaptations and variation in human organization and has tested this hypothesis on the basis of the recent bison-hunting groups on the Great Plains. The following chapters turn to one specific portion of the Plains—the Southern High Plains of west Texas and eastern New Mexico—to consider the relationship between changing ungulate adaptations and human organization during the Paleoindian Period (12,000 to 8000 B.P.) in light of this hypothesis. This chapter first describes the modern environment of this portion of the Plains, emphasizing those aspects of the environment that are particularly relevant to the issues that are important to the present study. Chapter 5 describes the modern climate of the region, and this discussion is not repeated here. This chapter then summarizes the data on climatic and other environmental conditions during the late Pleistocene and early Holocene and discusses the implications of these conditions for ungulate, particularly bison, populations in the region.

1. PHYSIOGRAPHY

For the purposes of this study, the Southern High Plains are defined as the region from the Canadian River on the north to a line extending eastward from the corner of the Texas/New Mexico border toward Midland, Texas, on the

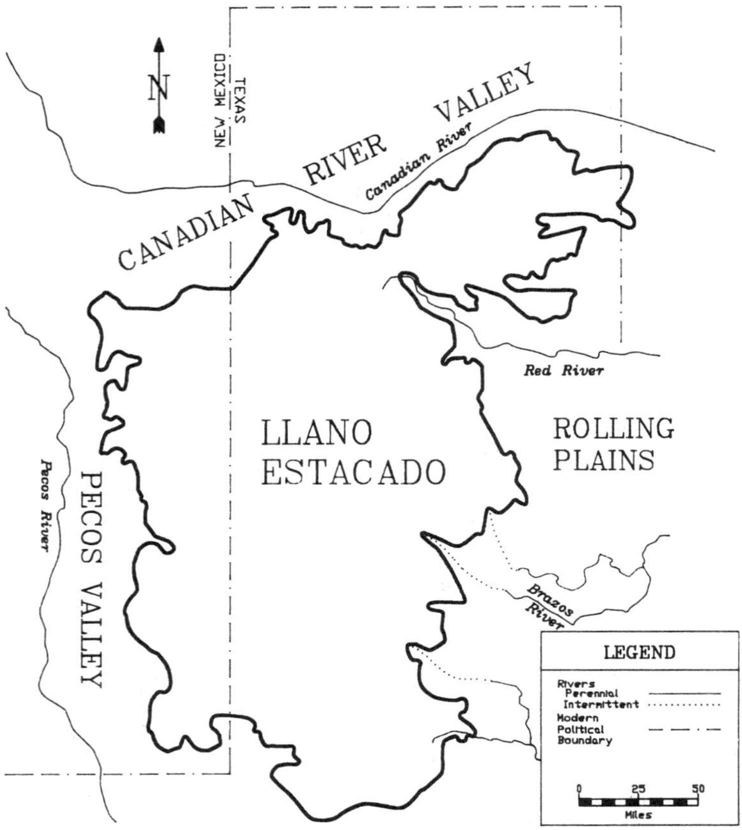

Figure 9-1. Physiographic provinces of the Southern High Plains.

south, and from the Pecos River on the west to the western portion of the Rolling Plains on the east (Figure 9-1). This region includes an area of approximately 120,000 square kilometers. It is conventionally divided into four relatively distinct physiographic provinces: the Canadian River Valley on the north, the Pecos River Valley and adjacent Mescalero Pediment on the west, the Llano Estacado, or Staked Plains in the center, and the Rolling Plains on the east. Figure 9-1 shows these four regions; Figure 9-2 shows the study area in somewhat more detail.

The Llano Estacado, the largest portion of the study region, is a huge, flat plateau that is bounded on the west, north, and east by cliffs that reach heights of 300 meters in the north along the edge of the Canadian River Valley and are lower to the south; on its southern edge, it grades into the Edwards Plateau. The surface of the plateau slopes from an altitude of 1,500 meters in the northwest to

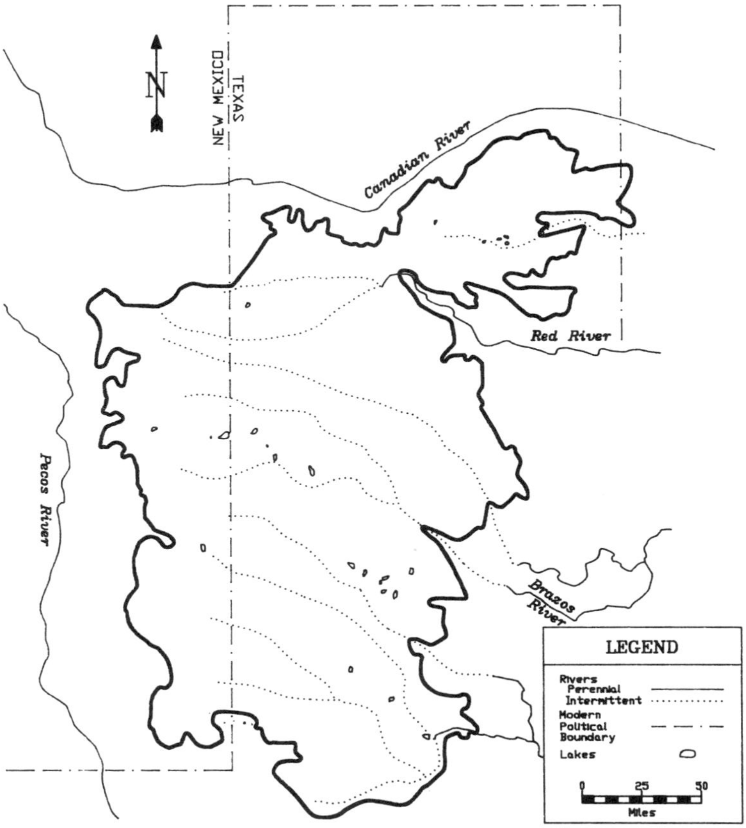

Figure 9-2. The Southern High Plains.

750 meters in the southeast, a drop of 1 to 2 meters per kilometer. The most impressive aspect of the Llano Estacado is its almost complete lack of relief: in most areas, there are no natural features more than 10 or 20 meters high to break the apparently endless monotony of the Plains.

However, this province is not simply a featureless plain. Topographic variability does exist, particularly near water. There are three principal sources of water on the Staked Plains: a series of drainages, or draws, running from west to east and emptying out onto the Rolling Plains; approximately 30 pluvial lakes; and literally thousands of small deflation basins that hold water after seasonal rains. Of these sources, only the downstream portions of the drainages and a few of the lakes are permanently wet at present.

The draws are the relics of an ancient drainage system whose flow was pirated during the Pleistocene by the Pecos River. At their eastern ends, these draws have incised deep canyons into the caprock escarpment; progressing west. they become steadily shallower. Their route generally varies between relatively straight segments and sharp meander bends that are entrenched into the underlying bedrock (Holliday 1983; Stafford 1981). The only supplies of water for these streams and for all other water sources on the Llano are local rainfall and the regional aquifer. The steep cliffs surrounding the Llano have cut off all outside sources of water, and the resulting hydrologic isolation has made the area an attractive "natural laboratory" for studying environmental change (cf. Wendorf 1975a:3).

The pluvial lakes of the Southern High Plains have been studied intensively by Reeves (1965a, 1965b, 1966a). The usage of the term *pluvial lakes* here is intended to distinguish between these large lake basins, which were formed mainly by fluvial processes during pluvial periods in the past, and smaller lake basins in the region, which were formed almost entirely by deflation during arid periods. All of these large basins appear to have existed for at least the past 20,000 years (Reeves 1966a), and they range in size from slightly less than 4 to more than 30 square miles. Most of them now contain water only seasonally, and there are well-developed playas visible in them during most of the year.

These lakes have extensive dunes forming massive ridges on their northern and eastern edges. These dunes fall into three groups by age (Reeves 1965b; Melton 1940): those farthest from the modern playas were formed before 15,000 B.P., a much larger series primarily on the eastern edge of the playas was formed between 15,000 and 5000 B.P., and small active dunes on the northern and northeastern edges of the playas were formed during the last 5,000 years.

Thousands of small deflation basins scattered over the otherwise unwatered uplands provide a third source of water in this province. These basins range in size from one quarter to one mile in diameter and may be from 20 to 30 feet deep. Like most of the pluvial lakes, they hold water for a time after heavy rains.

The other major topographic feature of the Llano Estacado is a massive dune belt running from east to west along Blackwater Draw. Most of these dunes appear to have formed during the mid-to-late Holocene: Haynes (1975) found only very localized dune deposits contemporary with or older than the Paleoindian Period.

Off the Llano Estacado, the appearance of the region has been extensively modified by erosion. On the east, the headward erosion of the draws over time has formed an escarpment several hundred feet high, marking the boundary between the Llano and the adjacent Rolling Plains. This escarpment marks a zone ranging in width from 1 to 6 miles that is characterized by a badlands topography with steep slopes, isolated mesas and peaks, and narrow, V-shaped drainages (Fenneman 1931; Gould 1906). Farther east, this topography gives way to the true Rolling Plains, a region of gently undulating topography with occasional sandstone hills rising as much as 100 feet above their surroundings.

The Rolling Plains are drained to the east by the Red, Brazos, and Colorado rivers, which are fed by the draws on the Llano Estacado and run in sandy beds cut 100 to 200 feet below the adjacent plains in valleys several miles wide (Gould 1906). These rivers are perennial, and in the past were fed not only by runoff from the west but by streams originating from numerous springs along the caprock escarpment that took water from the regional aquifer. Recent overuse of this aquifer, however, has dried up many of these springs (Speer 1980).

In the north, the Canadian River and its tributaries have eroded a badlands similar to that found east of the Llano Estacado. The river runs in a broad sandy channel as much as 1,000 feet below the High Plains surface; the valley of the Canadian ranges from 5 to 20 miles wide at the top. The region between the river and the adjacent Plains, the "Canadian Breaks" (Kennard 1975), is extremely rugged, with steep mesas, sharp canyon rims, and deep canyons; access to the plains from the breaks is possible in only a few places. Streambeds in this region are narrow and steep-sided. The Canadian River originates in the Rocky Mountains of northeastern New Mexico, where it is fed by mountain springs and local runoff. Additional water runs into the river from local tributaries carrying water from springs along the High Plains escarpments bordering the valley as well as from limited local runoff. Its flow is uncertain, varying from a complete absence of water on the surface to a torrent, but water can almost always be reached by digging even when the river is dry (Gould 1906).

The topography of the other major river valley in the study area, that of the Pecos, is radically different. The Pecos River runs in a broad, flat valley between two wide plains that cover much of eastern New Mexico (Reeves 1976). The river is fed primarily by drainage from the Sangre de Cristo Mountains where it originates and by numerous streams draining from the Sacramento and Sierra Blanca mountains to the west. All of these tributaries are now ephemeral. This study is concerned with the eastern portion of the Middle Pecos Valley and the southern portion of the Upper Pecos Valley as they are defined by Jelinek (1967). This region lies east of the Pecos River and extends from the Texas/New Mexico border on the south to just south of U.S. Highway 66 on the north. It is divisible geologically into the area immediately adjacent to the river and the area between the river and the Llano Estacado.

The Pecos River itself runs in an alluvium-filled channel, bordered by a low bluff on its east edge. The area along the river includes three distinct terraces (Fiedler and Nye 1933). These are known as the Blackdom Terrace, formed during the middle Pleistocene, the Orchard Park Terrace, formed during the late Pleistocene, and the Lakewood Terrace, formed during the Holocene. The Pecos has cut its present channel to a depth varying from 10 to 30 feet in the last of the three terraces. The Orchard Park Terrace is 5 to 10 feet above the Lakewood Terrace, and the Blackdom Terrace is 30 to 50 feet higher yet. Southeast of Roswell, a series of solution/subsidence cavities that hold permanent water have formed along the base of the bluffs on the east side of the river.

Between the river and the Llano Estacado is a broad, flat plain known as the Mescalero Pediment. The plain is separated from the High Plains proper by an escarpment that grades from a steep cliff in the north to a low, dune-covered slope in the south known as the Mescalero Ridge. Like the Llano Estacado, the Mescalero Pediment is hydrologically isolated, receiving water only from local precipitation. In this respect, it contrasts with a similar pediment to the west of the river that receives runoff from the mountains to the west. A series of small ephemeral streams run across this plain from the edge of the Llano to the Pecos. A few relatively large lake basins and many smaller sinkholes and deflation basins can be found in this part of the Pecos Valley, although they are not as numerous as those on the Llano and have not been studied as extensively. Most of the surface of the Mescalero Pediment is mantled with shifting sand, particularly in the southern portion of the study area (Fiedler and Nye 1933; Green 1961; Nials 1980).

2. VEGETATION

In the modern classification of North American grasslands, the study area falls into the Shortgrass Prairie region, although Allred and Mitchell (1955;10) refer to it as Mixed Prairie. The major species in the region are short and midgrasses, dominated by buffalo grass, blue grama, western wheatgrass, sand dropseed (*Sporobolus cryptandrus*), and threeawn (*Aristida* sp.), along with prickly pear (*Opuntia* sp.). This association grades into Desert Plains Grassland dominated by buffalo grass, galleta, and black grama in the southern Pecos Valley. In canyon breaks and sandy regions where better growing conditions for grasses are found, the High Plains Bluestem community (Allred and Mitchell 1955:10–11) occurs. This association includes taller eastern grasses, particularly little bluestem, switchgrass, sideoats grama, and indian grass (*Sorghastrum nutans*).

The major additions to these grassland associations on the uplands occur near water, where occasional trees and bushes such as hackberry (*Celtis* sp.), willow (*Salix nigra*), and cottonwood (*Populus sargentil*) can be found. Early settlers on the Llano Estacado particularly noted extremely dense growths of tall grasses in and around the small playas there, although most of these concentrations have now been grazed off. In the southern parts of the region, dense groves of mesquite (*Prospis glandulosa*) can be found in similar contexts. Saltbush (*Atriplex canescens*) and other salt-resistant species can also be found adjacent to the large saline lakes in the region. In areas of deep sand, particularly on the southern half of the Llano Estacado and the Pecos Valley, the High Plains Bluestem association is accompanied by fairly dense stands of shin oak (*Quercus havardii*) and sand sage (*Artimesia filifolia*). Particularly in the north, the sheltered areas below the caprock escarpment are lightly wooded with several species of juniper (*Juniperus pinchotii* and *J. monosperma*), together with shin oak, mesquite, and cholla cactus (*Opuntia imbricata*), and Wendorf (1975a:9) notes that historical accounts mention pinyon (*Pinus edulis*) in these areas.

The bottomlands of the Canadian River and its tributaries support a far more abundant growth of plants. Trees found there include cottonwood, willow (*Salix* sp.), elm (*Ulmus americana*), ash (*Fraxinus texenis*), and oak (*Quercus* sp.). A variety of berries, among them grapes (*Vitus acerifolia*), grow in the underbrush, dense growths of tall grasses are found in the open river and stream bottoms, and the immediate edges of the river channel support such water-loving species as cattails (*Typha latifolia*). The Spanish who visited the valley in the late 1500s explicitly noted the abundance of wild fruits there (Hammond and Rey 1966), mentioning plums (*Prunus gracilis*), persimmon (*Diospyros virginiana*), chinaberry (*Melia azedarach*), and soapberry (*Sapindus drummondii*). Overgrazing in the past 100 years has greatly increased the extent of mesquite and cholla in the region at the expense of the grassland; in the past, mesquite was mainly found on the edges of the valley (Marmaduke and Whitsett 1975:73–74).

The flora in the Rolling Plains differs from that elsewhere by the addition of substantial amounts of mesquite on the uplands. Blair (1950) refers to this region as the Mesquite Plains, and McDaniel (1978) discusses its vegetation in considerable detail. In dissected areas near drainages, the mesquite grassland is replaced by the Juniper River Breaks vegetation, mainly distinguished by an abundant growth of juniper. Immediately adjacent to watercourses is a riparian zone with salt cedar (*Tamarix gallica*), a recent European introduction, present on major streams and absent elsewhere. Major streams support a variety of trees and shrubs, including cottonwood, hackberry, oak, artemisia (*Artemisia filifolia*), and a variety of berries. Lesser drainages support a wider variety of trees and shrubs, adding elm, pecan (*Carya illinoisensis*), ash, and plum, among others. Both types of drainages also support dense growths of grasses, generally mid- to tall species near major streams and short to mid- species near minor streams.

The Desert Plains Grasslands in the southwest portion of the study area, in the southern portions of the Middle Pecos Valley, also show a distinct vegetation, with sparser short grasses associated with creosote (*Larrea tridentata*) and blackbush (*Flourencia cernua*) in areas without dune cover and with mesquite, shin oak, and sandsage in the dunes. In the most southerly areas, grassy areas are only occasionally present, and these other species are dominant (Nials 1980). Much of the spread of this shrub cover at the expense of the grasslands has occurred in the past 50 years (Hennessy *et al.* 1983).

3. A GENERAL BASIS FOR RECONSTRUCTING PAST ENVIRONMENTS

The available paleoenvironmental information discussed in the next chapter is not uniformly distributed across the study area. Detailed research has been confined largely to the Llano Estacado; much less work has been conducted in the Pecos Valley, the Canadian River Valley, and the Rolling Plains. Reconstructions of environmental conditions in these latter areas must therefore necessarily be less reliable than reconstructions of the Staked Plains. However, this problem can be mitigated by considering the overall pattern of environmental variation in the region, particularly the pattern of variation in vegetation.

The general differences in vegetation in the four provinces reflect the differences in the amount of moisture available in them, which is controlled in turn both by the general physiography of North America and by the specific physiography of the Southern High Plains. First, the east-to-west decrease in precipitation across the Great Plains and hence across the study area is caused largely by the rain-shadow effect of the Rocky Mountains in the west and must therefore have been characteristic of the Plains throughout at least the Holocene. Second, as Wendorf (1975a:3) notes, the Llano Estacado subsection of the study area is hydrologically isolated, receiving water only from local rainfall. It has not been widely noted, however, that this is also true for that portion of the Pecos Valley between the Pecos and the caprock escarpment, which receives only limited runoff from the Llano as a supplement to precipitation. Most of the runoff from the Staked Plains flows into the lakes in the region or down the draws and out onto the Rolling Plains. The Canadian and Pecos Rivers receive water from the mountains and plateaus of northeastern New Mexico, but the flow in the latter river is supplemented by fewer tributaries than that in the former river, and most of these tributaries run from the west, outside of the region considered here.

These patterns have several important implications for this analysis. First, the hydrologic isolation of the region except for the areas immediately adjacent to the Pecos and Canadian Rivers implies that water-related environmental conditions in the Pecos Valley, Llano Estacado, and Rolling Plains should respond in more or less similar ways to local climatic patterns. As the next sections of this chapter discuss, this assumption is supported by the similarity of early Holocene alluvial stratigraphy throughout the region.

Furthermore, relative differences in water and other resource abundance between the four provinces should have persisted throughout time: the Canadian Valley and Rolling Plains should always have been wetter and better vegetated than the Pecos Valley and Llano Estacado. The east-to-west decrease in rainfall implies that of these two, the Llano Estacado would have been better-watered than the Pecos Valley. In addition, the north-to-south increase in temperature found in the area today must have been present throughout prehistory.

4. PALEOINDIAN PERIOD CLIMATIC CHANGE

The presence of large lake basins on the Llano Estacado clearly implies that conditions at some time in the past were considerably wetter than they are at present. Reeves (1966a; Reeves and Parry 1965) has shown that these lakes were last full during the Tahoka Pluvial, a period dating between 20,000 and 14,000 B.P. (also see Wendorf 1975b:261–264). Integrating data on the size of the lakes, Pleistocene snow lines in New Mexico, and evaporation and runoff rates, Reeves (1965a, 1965b, 1966b) argues that high water levels in the lakes were probably maintained by a combination of increased precipitation and decreased temperature. He estimates that annual precipitation was as high as 85 to 88 centimeters (33 to 34 inches) and that mean annual temperatures were 9 °F cooler than at present, with most of the cooling due to a reduction of 18° in summer temperatures. Smith and Smith-Perrott (1983:205) suggest that groundwater may have made more of a contribution to these lakes than Reeves states, but total precipitation must still have been significantly higher than in recent times. Furthermore, Pleistocene precipitation was apparently much more evenly distributed throughout the year than is modern precipitation, with winter precipitation particularly increased (Reeves 1966b:644).

The most obvious climatic trend after the Tahoka Pluvial is a progressive decrease in effective moisture, probably because of decreasing precipitation and increasing temperatures. The stratigraphic sequences in a series of sites in the draws in the region document this trend, showing a shift from free-flowing streams between approximately 12,000 or 13,000 and 11,000 B.P., to open ponds and lakes between 11,000 and 10,000 B.P., and finally, to an aggrading marsh with little or no open water from 10,000 to 6300 B.P.; the upper portions of these marsh deposits often show signs of erosion (Haynes 1975; Holliday 1985; Holliday *et al.* 1983; Honea 1980; Johnson *et al.* 1982; Stafford 1981). Evidence of extreme aridity after 6000 B.P. is shown at Blackwater Draw (Evans 1951) and Rattlesnake Draw (Smith *et al.* 1966) in eastern New Mexico and at Marks Beach in Texas (Honea 1980), where wells dated between 6000 and 5000 B.P. were dug down to a buried aquifer in locations that had previously contained standing water.

There is also some evidence for fluctuations in this progressive trend. The stratigraphic record at Blackwater Draw Locality Number 1 (Haynes 1975) indicates that the water table stood at the surface of the draw before approximately 13,000 B.P., dropped substantially at 12,000 B.P., rose to its former level until 10,000 B.P., dropped again until 9000 B.P., rose again, although not to its previous maximum, at about 8000 B.P., and steadily dropped thereafter. Evidence for these fluctuations is much less pronounced in the strata farther to the southeast, particularly those at Lubbock Lake (Holliday 1983; Stafford 1981), probably because the water table is higher in the east due to the southeasterly slope of the

Llano and because the downstream portions of the draws capture runoff from a larger area than the upstream portions.

The increasing seasonality of post-Tahoka precipitation is also documented by an analysis of snails at the Lake Theo site, on the western edge of the Rolling Plains (Johnson *et al.* 1982). This analysis indicates higher winter precipitation and more evenly distributed summer precipitation than at present between approximately 11,000 and 10,000 B.P. This study further shows evidence for cooler summer temperatures than at present. Similarly, the fauna from the Lubbock Lake site indicate a trend toward increasingly colder winters and warmer summers throughout the early Holocene (Johnson 1988a).

These trends fit well with evidence for late Pleistocene and early Holocene climates in much of central North America. Bryant and Shafer (1977), for example, note increasing aridity in much of Texas during this time, and Wright (1970) and McMillan and Klippel (1981) find evidence for a similar pattern on the Central Plains and adjacent prairies. The increasingly seasonal distribution of rainfall and the trend toward more extreme seasonal differences in temperature are also typical of post-Pleistocene climatic shifts throughout the world (Graham and Lundelius 1984; Guthrie 1984). Evidence for year-to-year variation in climate is obviously difficult to obtain for any prehistoric period, but Hershfeld (1962) has shown that variation in precipitation increases as the number of days on which precipitation falls decreases. This variation should then have increased as precipitation became more and more concentrated in the summer and less evenly distributed even during this season.

5. AVAILABLE SURFACE WATER

As was noted before, the stratigraphic record in the draws on the Llano Estacado indicates that there was a progressive decrease in the amount of water flowing through the region during the early Holocene. Geologic research in entrenched meander bends of the draws, such as that at Lubbock Lake (Holliday 1983; Stafford 1981), indicates that there was a shift from running water to alternating open ponds and marshes after 11,000 B.P. and then to permanent marshes with little or no open water after 10,000 B.P. Free flowing streams persisted in the more open straight portions of the draws until this later date (Holliday 1985).

There is also evidence that the larger lakes remained wet throughout the period considered here, although they obviously contained much less water than in earlier times. Reeves (1965a) notes the presence of remnants of post-Tahoka lacustrine deposits under the dunes fringing the present playas in the basins, and Haynes (1975:77–81) notes strata in Rich Lake in Terry County, Texas, which are upland facies of lake or pond sediments and which date between 12,000 and 6000 B.P.

In addition, there appear to have been far fewer small deflation basins on the upland portions of the Llano Estacado than are found there at present. Reeves and Parry (1969) identify three classes of small deflation basins in this area: small, circular basins that lack fringing dunes, larger basins with their eastern edge straightened and with dunes on their eastern flanks, and basins with both their eastern and western edges straightened that also have dunes to the east.

Changes in basin shape such as these occur when they contained water for extended periods of time during wetter periods in the past: basins that are full are lengthened perpendicularly to the direction of the prevailing wind through end-current erosion and first straighten their lee edge (here, the east) and then their windward edge (here, the west). The three distinct shape classes indicate three

Table 9-1. Frequencies of Different Classes of Small Deflation Basins on the Llano Estacado by USGS 7.5-Minute Quadrangle

Quadrangle	Class 1[a]	Class 2[b]	Class 3[c]	Unknown	Total
Texas					
Amherst	10	7	8	5	30
Baileyboro	1	1	14	0	16
Brownfield West	6	5	6	1	18
Buffalo Stadium	5	6	5	1	17
Dimmitt	11	12	1	2	26
Hale Center	6	16	19	4	45
Lamesa South	14	8	2	0	24
Lehman	5	10	5	2	22
Levelland West	7	21	6	1	35
New Deal	5	7	8	3	23
Parmerton	6	6	4	1	17
Plains	1	0	0	0	1
Tahoka	5	7	4	1	17
Tule Lake	3	9	11	1	24
Westway Northwest	12	2	1	0	15
New Mexico					
Broadview	6	7	5	1	19
Dora Northwest	8	8	5	0	21
Hobbs West	3	4	17	2	26
Melrose East	2	1	5	0	8
Milnesand Northwest	2	2	0	0	4
Total	118	139	126	25	408

[a]Class 1: Vigo Park.
[b]Class 2: Altithermal.
[c]Class 3: Recent.

distinct periods of basin formation, and Reeves and Parry correlate them with periods of regional deflation during, respectively, Recent times, the Altithermal (6500 to 4500 B.P.), and the Vigo Park Interval (17,000 to 16,000 B.P.), an arid period within the Tahoka Pluvial.

Although Reeves and Parry did not attempt to estimate the number of basins formed during each of these periods, application of their criteria to the small lake basins on 21 USGS 7.5-minute maps distributed over the Llano Estacado indicates the frequency of the three shape classes. Two-thirds of the basins found at present appear to have formed during Altithermal or Recent times, after the end of the Paleoindian Period (Table 9-1), implying that surface water was much less widely available at that time than it is now.

Overall, the volume of permanent surface water on the Southern High Plains declined during the early Holocene. The locations of this water (in the draws and lakes), though, probably remained constant. Because there were many fewer deflation basins on the uplands at this time, there were also many fewer locations where seasonal water could be found away from these permanent sources. One effect of an increasingly seasonal pattern of precipitation must also have been a progressive restriction in the proportion of the year during which the existing upland deflation basins held water.

6. VEGETATION AND FORAGE PRODUCTION

There are two distinct reconstructions of the vegetation on the Southern High Plains during this period. The first of these, based almost entirely on pollen evidence, asserts that following the arid period at approximately 12,000 B.P., during which the pollen indicate essentially modern vegetation, there was an extensive cover of pine and spruce on the Llano Estacado. This is inferred on the basis of extremely high frequencies of pine pollen in samples taken from a number of localities. Oldfield and Schoenwetter (1975:167) reconstruct this cover as "open patches of Ponderosa pine parkland, with some admixture of spruce. . . . At lower altitudes, there was probably mixed pine woodland covering less of the area, and hardly any spruce." Wendorf (1970:29) characterizes this vegetation as "a dense boreal pine forest with occasional spruce." After 10,000 B.P., these authors see a return to a grassland.

Other researchers, however, find no evidence for pine forests of any kind, although there was probably some pine present in more sheltered areas and near water. Reeves (1976:223) summarizes this research: "Contrary to some interpretations made wholly on pollen evidence, . . . no stratigraphic, pedologic, or paleontologic evidence exists that the Southern High Plains was ever heavily forested at any time during the Quaternary." Preliminary soils research conducted by Haynes (1975:76) as part of the same project that produced the pollen results, in fact, suggests that the basic vegetation of the region was an open grassland throughout the terminal Pleistocene and early Holocene. Most recently, extensive analyses of the soils and fauna at the Lubbock Lake site (Holliday *et al.* 1983; Johnson 1988a) have uniformly found evidence for essentially open grasslands throughout this period.

This last research also suggests an explanation for the earlier pollen results. Fluctuations in the water table in the region have extensively oxidized the early Holocene deposits there, and Stafford (1981:559) reports pollen counts ranging from 10 to 1000 grains per gram in these deposits at Lubbock Lake. In contrast, counts in unoxidized downstream segments of the same strata range from 1,000 to 300,000 grains per gram (McLaughlin n.d.). The implication that as much as 99% of the pollen in these deposits may have been destroyed suggests that the earlier palynological results are extremely suspect.

Although the relative resistance of many pollen types to destruction by oxidation is unknown, pine pollen is among the most resistant of the species that have been studied (Hills 1969:74–76). Of the other types of pollen noted in Paleoindian deposits in the study area, those whose resistance is known (such as *Betula, Alnus, Acer, Typha,* and *Ulmus*) are uniformly present only in trace amounts (Oldfield 1975; Schoenwetter 1975) and are uniformly more susceptible to oxidation than pine. These data suggest very strongly that the dominance of pine pollen found by past palynologists in the region is the extreme result of differential pollen preservation and that this dominance provides no adequate basis for inferring the presence of an extensive pine forest on the Southern High Plains at any time during the late Pleistocene or the early Holocene (cf. Holliday *et al.* 1985: Holliday 1986).

Rather, the major shifts in vegetation on the Llano Estacado indicated by the faunal and soils evidence are from a parkland or savanna dominated by mixed grasses with "isolated clumps of hackberry and other deciduous trees" before 11,000 B.P. to an open grassland with decreasing amounts of brush and trees remaining near water thereafter (Holliday *et al.* 1983:172, 174–175; Johnson 1988a). The open topography and hydrologic isolation of the Pecos Valley imply that similar conditions existed there. The snail data from the Lake Theo site on the western Rolling Plains (Johnson *et al.* 1982) suggest that denser woodlands existed in the more broken terrain below the caprock esacarpment and, by extension, in similar regions in the Canadian River Valley.

The apparent presence of a grassland habitat in the region throughout the period considered here makes it possible to assess the effects of the changing climate on late Pleistocene and early Holocene patterns of forage production there. Decreasing precipitation would progressively have reduced total forage production, and increasing variability in precipitation from year to year would have steadily increased annual fluctuations in forage conditions. As precipitation became more and more concentrated in the summer months, the amount and quality of forage produced during other seasons of the year would also have been reduced, and this trend would have been exacerbated by changing temperatures. Increasingly colder winters would have further inhibited grass growth during this season, and increasingly warmer summers would have increased the proportion of warm-season species in the region, leading to shorter periods of maximum forage production beginning later in the year and to more erratic patterns of growth throughout the year.

7. UNGULATE ADAPTATIONS

Two well-known and important changes occurred in the fauna of the Southern High Plains during the terminal Pleistocene and early Holocene. The first of these was the extinction of most of the large species of mammals in the region, which apparently occurred between 11,000 and 10,000 B.P. (Meltzer and Mead 1983). The second was a major set of changes in the morphology and presumably the adaptation of the bison, the dominant remaining large ungulate, which occurred progressively throughout the early Holocene. This section discusses the effects of these changes on the structure of the faunal resource base in the region.

Although explaining the causes of the late Pleistocene extinctions in North America is not the goal of this study, existing explanations have important implications for terminal Pleistocene human adaptations. Scholars who have considered this event generally take two positions. The first, referred to here as the overkill theory, relates these extinctions to the appearance of human beings in the New World just after 11,000 B.P. This position notes that the area covered by grasslands was expanding at the end of the Pleistocene and concludes that large herbivore populations should also have been expanding. The only factor that is seen to account for the fact that they did not expand is human predation, and the overkill theory holds that human hunters entered the New World and simply slaughtered everything in sight (Martin 1973, 1984). The second position holds that the late Pleistocene extinctions occurred because of a complex set of changes in the environment, probably including the appearance of human hunters (Graham and Lundelius 1984; Guthrie 1984; Kiltie 1984).

This second position is favored here for two reasons. First, there is simply no archaeological evidence that the earliest human hunters preyed frequently on any species of large mammal in North America other than the mammoth. Butchered remains of other large species including bison, camel, and horse are known from several sites (Frison *et al.* 1977; Hemmings 1970; Johnson 1977), but these remains usually consist of only a few skeletal elements. Martin (1984:359–360; Mosiman and Martin 1975) attempts to explain this fact by the concept of "blitzkrieg," arguing that human hunters expanded so fast that the period during which they overlapped with the extinct fauna was too short to leave extensive archaeological traces. However, this does not account for the relative abundance of elephant kill sites or for these hunters' failure to wipe out the North American bison. Only further field investigations can truly test the blitzkrieg notion (Martin 1984:396–397), but, at present, it is simply too facile a rationalization of a serious empirical problem with the overkill hypothesis.

Possibly more important is the extreme oversimplication of the nature of the environmental changes that occurred at the Pleistocene/Holocene boundary on which the overkill theory rests. Contrary to assertions that similar changes occurred repeatedly during the Pleistocene without producing mass extinctions (Martin 1984:357), Graham and Lundelius (1984), and Guthrie (1984) have shown that the Pleistocene/Holocene transition was unique, involving a fundamental restructuring of the kinds and distributions of plant foods and a major change in the seasonal pattern of forage availability. Chapters 3 and 4 have shown how closely climate, forage production, and ungulate adaptations are linked, implying that changes such as these could well have had catastrophic effects on the resident ungulate populations. The ecological naivete of the overkill theory makes it unacceptable in its present form.

Regardless of what caused the late Pleistocene extinctions, the disappearance of most of the species of large ungulates on the Southern High Plains and predictable effects on the structure of the region's faunal resource base. As Chapter 4 discusses, diverse species of herbivores competing for forage and water in a single region tend to evolve feeding strategies specialized on different species of plants and different parts of the same species of plants. Specialization on different plants that live in different environmental zones tends to keep species spatially separated, and specialization on different parts of the same species of plants leads different species of herbivores to feed in a given local area at different times: a species that prefers leaves may feed in an area first, exposing the stems for later feeding by species that prefer them. Such preferences create a regular succession of animals moving through an area, and such a succession must have existed on the North American grasslands during the Pleistocene (Graham and Lundelius 1984:227–229).

One effect of such a succession for a hunter looking for game is to increase the likelihood that at least some species of animal will be in a given local area at a given time. A diverse fauna thus tends to keep animals relatively evenly spread across the landscape, a pattern that was probably enhanced by the higher populations of herbivores resulting from higher forage production and the ability of a diverse fauna to use this production more completely than a fauna dominated by a single species. The tendency for large mammals to avoid each other at waterholes would also have tended to spread animals out across the landscape. Relative to later times, then, the late Pleistocene fauna were abundant and dispersed. Low seasonality, a relative lack of year-to-year variation in climatic conditions, and low summer temperatures would also have kept forage production high for more of the year, similar from year to year, and regular within a given year. This would have minimized seasonal changes in herd size and kept herd movements regular and repetitive from year to year and within a season.

The bison was the only major game species remaining in the study area after the rest of the late Pleistocene megafauna had disappeared, and there are major changes in bison morphology after this time. The existing data provide a relatively clear picture of the nature of these changes, and the remainder of this section considers their relation to the environmental conditions outlined before and the evidence they provide about changes in the nature of bison adaptations on the Southern High Plains during the early Holocene.

Much of the recent debate over the nature of early Holocene bison evolution (McDonald 1981:38–131; Walker and Boyce 1985; Wilson 1969) is about how many different species of bison should be recognized and what these species should be named rather than about the adaptive or ecological meaning of the observed changes. This disagreement exists at least partly because the available data indicate that these changes occurred gradually over several thousand years, and, in essence, this debate is therefore about how to partition an evolutionary continuum. In order to move beyond taxonomy and understand how bison adaptations were changing during the early Holocene, it is the continuum itself that is important and that is the emphasis of the present discussion.

The most obvious trend in bison morphology over this period is a continuous and substantial decrease in overall size (Bedord 1974; Hughes 1978; McDonald 1981), a trend that the species shares with most of the large herbivores in North America (Guthrie 1984). Wheat (1972:114) estimates that early Holocene bison produced about 25% more meat than modern bison. At the same time that the bison were shrinking, their horns were changing their shape, the structure of their frontal bones and the orientation of the orbits and the frontals relative to the spine were changing, the head was being lowered, the legs were became relatively slimmer overall, and the proportions of some of the individual bones in the legs became mechanically more adapted to power than to speed (Guthrie 1980; McDonald 1981; Smiley 1978; Wilson 1978). Sexual dimorphism also appears to be greater among Pleistocene than modern bison (Guthrie 1980; McDonald 1981:151).

McDonald's data (1981:178, 191) also indicate that the legs of the bison became longer in proportion to the overall size of the body, although he does not note this trend. Table 9-2 presents mean values for skull volume and for the total length of the bones in the front and rear legs for bison that McDonald (1981:57–107) classifies as *Bison latifrons, Bison antiquus antiquus, Bison antiquus occidentalis,* and *Bison bison bison*. Whether or not these taxa should be considered species or subspecies (cf. Walker and Boyce 1985), they represent a temporal sequence from mid-Pleistocene to Recent times, with *Bison latifrons* being the oldest and *Bison bison bison* the youngest.

Table 9-2. Late Pleistocene/Early Holocene Bison Leg Lengths (Millimeters) and Skull Volumes (Centimeters Cubed)[a]

Species	Front leg	Rear leg	Skull	Front/skull	Rear/skull
Bison latrifons	1,033.5	1,243.4	7,592.2	0.14	0.16
Bison antiquus occidentalis	927.0	1,114.9	5,190.9	0.18	0.21
Bison antiquus antiquus	920.5	1,077.5	4,025.4	0.23	0.27
Bison bison bison	852.7	1,007.8	3,371.7	0.25	0.30

[a]From McDonald 1981:178, 191.

Many of the characteristics of fossil bison that are commonly measured, particularly measurements of the horns, appear to have changed in response to both the overall dwarfing trend through the late Pleistocene and early Holocene and to other factors (see later text). However, skull volume is likely to be a reasonably accurate measure of overall body size (cf. McDonald 1981:177). On this assumption, Table 9-2 also shows the ratio of total leg length to skull volume. This ratio increases steadily over time, indicating that the length of the legs relative to the overall size of the body increased, although their absolute length decreased as the bison became smaller overall.

Several hypotheses have attempted to account for these changes by reference to a variety of factors. A major trend that these authors (Guthrie 1980; McDonald 1981; Smiley 1978) have suggested is toward larger and larger herds over time. This is indicated particularly by the changes in the skull and by the selection for more powerful animals indicated by the changes in the bones of the legs. Reproductive success for male bison is achieved by combat in which the competing animals clash their heads together; this clash is often followed by a hooking motion where they attempt to impale each other with their horns. The changes in the skull show a constant trend toward smaller horns, stronger frontals, and an orientation of the head on the spine that can bear the force of such a clash more effectively. Smaller horns tend to reduce the likelihood that combats will be fatal because they make it more difficult to impale a competitor, and the advantage of greater power in this kind of fight is obvious. These changes probably indicate that the frequency of combat between individual bison was increasing steadily, indicating that the bison were coming in contact with one another more often, which in turn appears to be good evidence for increasing herd sizes.

McDonald (1981:209) fairly plausibly links the changes in the orientation of the orbits to a shift from a Pleistocene mixed grazing and browsing diet to a Holocene diet based almost exclusively on grazing: the head is held lower to graze than to browse, and the orbits must be higher to watch for predators and maintain contact with the herd while feeding. Explanations for the increase in herd sizes and reduction in overall size, though, are more controversial.

McDonald (1981:243–258) links these changes in bison adaptations to a change in the pattern of bison mortality caused by the appearance in the New World of human beings as predators at the end of the Pleistocene, a link that grows out of the overkill theory of the late Pleistocene extinctions discussed before. In his reconstruction, the dominant Pleistocene pattern of mortality was density-dependent, meaning that mortality rates varied with population density, increasing as the population increased and decreasing as it decreased. McDonald characterizes the Pleistocene fauna as *K-selected,* living under relatively constant environmental conditions with population densities at or near the carrying capacity of the environment (also see Pianka 1970). With the appearance of human hunters, he argues that mortality shifted to a more catastrophic pattern in which death rates are independent of population size. He characterizes the post-Pleistocene fauna as more *r-selected,* living under more variable environmental conditions and having population densities well below the carrying capacity of the environment.

McDonald relates these supposed changes to the trends toward smaller bison and larger herds by appealing to Pianka's (1970) and Western's (1979) data indicating that such trends are associated with more rapid maturation rates and higher reproductive potential, characteristics that are adaptive under more *r*-selected conditions. His analysis of the frequencies of radiocarbon-dated occurrences of bison and other fauna indicates that the populations of all large ungulates were decreasing during the terminal Pleistocene and that this decrease continued for the bison into the early Holocene (McDonald 1981:244–245). Following the overkill theory, McDonald asserts that this should not have occurred because the grasslands were expanding and, he argues, ungulate populations should therefore have increased. Human hunting is the only force he sees in the environment that can account for the patterns in his data.

Unfortunately, the assumption that a larger area covered by grasslands necessarily would improve forage conditions for herbivores is incorrect. As Chapter 4 showed, the size of herbivore populations and the nutritional status of individual animals is controlled not by the total area producing food that a given species is capable of consuming but by the total amount of food produced within the area an individual animal is capable of exploiting and by the seasonal pattern of this production. Paleoclimatic data throughout the Plains indicate that effective moisture decreased during the late Pleistocene and early Holocene, and forage production per unit of area must therefore have decreased as well. Furthermore, the increasing seasonality of forage production during the early Holocene would have lengthened the period during which the nutritional value of the available food was below the maintenance requirements of the bison, probably depressing population sizes further.

Although McDonald's appeal to human predation does not appear to fit the facts, all of the observed changes can be linked directly to the changes in the regional pattern of forage production discussed before. This includes the increase in herd size just discussed, the dwarfing of the bison throughout the early Holocene, and the trend toward proportionately longer and slimmer legs.

First, Chapter 4 discussed the relationship between body size and forage conditions: as forage deteriorates, smaller animals have an advantage over larger animals because they require less food, need to spend less time eating, and therefore have more time available to seek out more nutritious plants. This is true whether the reduction in bison size is the result of genetic selection for smaller animals due to the reduced ability of larger animals to survive the winter or of a phenotypic shift caused by a reduction in the proportion of the year within which animals can find sufficient amounts of high-quality food to attain their maximum growth potential, as Guthrie (1984) suggests.

Second, an increase in herd size tends to increase the quality of the grasses in the region the herd inhabits by creating "grazing lawns" of densely packed, more nutritious plants that are kept in earlier stages of growth for more of the year. Larger herds thus could have partially counteracted the steady deterioration of forage conditions caused by decreasing precipitation and increasing temperatures.

Finally, the *proportionately* longer and slimmer legs of the more recent bison are almost certainly adaptations that improved the animals' ability to run efficiently for longer periods of time. This is particularly clear in the case of the length increase, which increases the length of the stride the animal can take. The shift toward more powerful animals noted previously came at the expense of their speed (Smiley 1978), but these other changes appear to have minimized the effects of this change on the animals' mobility. These apparently contrasting evolutionary trends probably reflect the conflicting structural demands of two selective forces: the direct reproductive advantage of victory in dominance battles, which requires strength, and the indirect reproductive advantage of being better-nourished, which requires the ability to obtain adequate forage.

As forage production declined and became more seasonal, animals that could search for forage over larger areas and could commute between water sources and relatively distant feeding areas would have been better nourished over the winter. The increasingly seasonal pattern of precipitation would also have reduced the proportion of the year during which the rainfall-dependent upland deflation basins could have been used as watering places, forcing the bison to use the more restricted permanent sources of water, and this would have occurred during the very seasons (fall, winter, and early spring) in which forage production would have been lowest. The forage around these sources would thus have been exhausted relatively quickly, further increasing herd mobility. Forage in any local area, in fact, would have been depleted more rapidly over time as herd sizes grew: although larger herds increase the quality of forage overall, they also consume local forage more quickly than smaller herds, forcing them to move more often.

The relatively restricted distribution of surface water except during the summer would also have tended to tie bison home ranges to permanent water and might have led the herds to abandon potential foraging areas that were too far from water (cf. Western 1975). Herd movements within these home ranges would also have become progressively more erratic over time. The relatively limited water available in the Pecos Valley away from the river itself and the minimal rainfall in that part of the study area should have made it less desirable for bison earlier than the rest of the Southern High Plains. Similarly, the most productive areas for the bison in the rest of the study area should have been toward the east, where higher rainfall increased forage production and greater numbers of permanent lakes and greater volumes of water in the drainages provided more reliable water supplies.

It is also important to note that the increase in herd size refers primarily to the maximum size of the herds present during the year, which was presumably attained during the rut. As forage production became more seasonal, forage conditions would have remained poor for more and more of the year, and the herds must have dispersed accordingly, forming their largest groups only during the summer when food was most abundant and nutritious. There must therefore have been a trend toward increasing seasonal differences in herd sizes during the terminal Pleistocene and early Holocene.

From an initial population of relatively larger, more solitary, and sedentary bison, decreasing forage production concentrated in a smaller and smaller portion of the year and distributed less and less predictably in space favored those animals who required less food, who tended to band together in larger herds whose collective grazing increased the quality of the food in their range, and who could move relatively efficiently over longer distances in search of food and water. The combination of simultaneous selection both for power over short distances and for running ability was probably the result of the competing reproductive advantages of strength in combat during the rut and better nutritional condition during the whole year. Animals lacking these characteristics would have tended to reproduce less often due to poor condition and less frequent victories in combat and would have been less well-nourished during the winter and therefore more likely to die of disease or starvation before the period of new grass growth in the spring.

8. SUMMARY

To summarize, forage production on the Southern High Plains during the terminal Pleistocene and early Holocene declined overall and became progressively concentrated in a shorter period of the year, spatially more and more erratic within a season, and more and more variable from year to year. Although the relation of these changes to the extinction of most of the Pleistocene megafauna is not clear, a major effect of these extinctions was to reduce the abundance and dispersion of the available animals in the region.

Changing forage conditions appear to account relatively well for the observed changes in bison morphology during the late Pleistocene and early Holocene. Individual bison became smaller and smaller but apparently lived in herds that became larger and larger and more and more mobile over time. Assuming that this mobility was tied to local patterns of forage production, as it is in all modern ungulates, the frequency, distance, and speed of herd movements must also have become more erratic within a season and less regular from year to year. In addition, the steady decrease in the length of the period of maximum forage production within a year would have led to increasing seasonal differences in herd size.

Finally, the direct link between forage production and ungulate population size implies that declining forage production must have steadily reduced the total number of animals in the region. A pattern of fewer animals aggregating into larger and larger groups also tends to reduce the number of locations where herds could be found and to increase the distance between herds. The herds should have been more concentrated in regions around permanent water, at least during the seasons when precipitation was low and the upland deflation basins were dry, and should have been more numerous farther to the east where forage production was higher.

In terms of the three basic variables defined in Chapter 2, total regional productivity generally decreased during the early Holocene as total forage production, and thus the number of animals in the region, decreased. Patchiness, the degree of clustering of the animals in space, increased as faunal diversity decreased at the end of the Pleistocene and as bison herd sizes subsequently increased. Constancy, the regularity of herd movements from season to season and from year to year, would have been high early in the Paleoindian period because ungulate movements and numbers changed less with the seasons and varied less from year to year. It would also have been easier to anticipate herd sizes and locations at this time because the animals moved less and in more regular patterns; contingency, the degree to which future herd sizes and locations could be predicted on the basis of present environmental conditions, would thus have been high.

As the extinctions left only the bison and as forage production progressively decreased, became more seasonal, and varied more from year to year, the herds must have become more mobile. Furthermore, their degree of mobility and their seasonal aggregations would have varied more from season to season and from year to year. Under these conditions, both constancy and contingency would have decreased fairly steadily. The implications of these changes for the early human occupants of the Southern High Plains are the subject of the next two chapters.

Chapter **10**

Paleoindian Adaptations on the Great Plains

This study emphasizes the relationship between the availability of animals on the Plains under different environmental conditions and the adaptations of the human beings preying on those animals. Such an emphasis presupposes a diet dominated by meat, and the specific hypothesis linking animal ecology and human adaptations proposed here presupposes a seasonal pattern of communal hunting. Considering the implications of this hypothesis for prehistoric societies requires evidence that these suppositions hold for the societies at issue. This chapter summarizes the chronological sequence of occupations on the Southern High Plains to provide a basic framework for a general discussion of Paleoindian adaptations on the Great Plains in this chapter and for the specific discussion of Paleoindian adaptations on the Southern High Plains in the chapter that follows. It then summarizes the evidence for Paleoindian subsistence, hunting, and aggregation patterns on the Great Plains.

1. CHRONOLOGICAL FRAMEWORK

Four relatively well-defined divisions within the Paleoindian Period are recognized on the Southern High Plains: Clovis (11,500 to 11,000 B.P.), Folsom (11,000 to some time between 10,500 and 10,000 B.P.), Plainview (dated to approximately 10,000 B.P. with exact beginning and ending dates unclear), and Firstview (post-Plainview to 8000 B.P.). These divisions are distinguished by changes in the types of projectile points found in the region and on the Plains in general and in the species of large mammals that these points were apparently

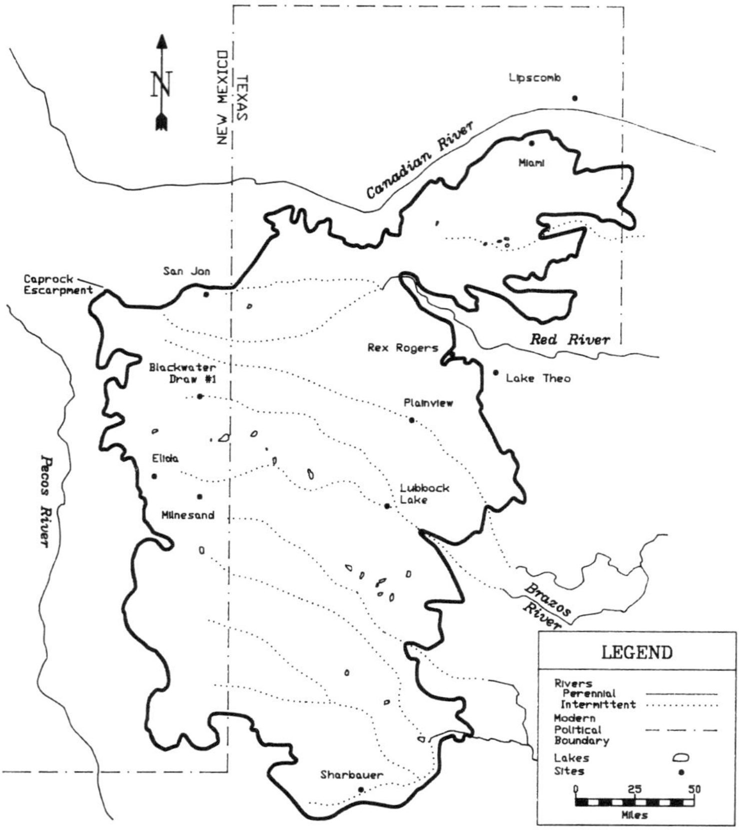

Figure 10-1. Locations of archaeological sites mentioned in the text.

used to kill. This section of the discussion emphasizes excavated, published sites; the following chapter considers Paleoindian sites that are recorded but not investigated. Figure 10-1 indicates the locations of the sites mentioned in this section.

The oldest certain human occupation known in the study area, as in the rest of North America, dates to the Clovis Period, between 11,500 and 11,000 B.P. (Haynes 1980; Holliday *et al.* 1983; Hughes and Willey 1978:25). Very few Clovis sites have been investigated systematically in the study area. Diagnostic projectile points and a relatively wide array of other tools associated with butchered mammoth and other fauna were dated to this period at Blackwater Draw in eastern New Mexico (Hester 1972), butchered mammoth and several points and a uniface were recovered at the Miami site in Roberts County, Texas (Sellards 1952:18–29), and butchered remains of mammoth, camel, horse, and several other species have been dated to this period at the Lubbock Lake site in Lubbock County, Texas (Holliday *et al.* 1983, 1985; Johnson 1988b).

Most of the large fauna on the Southern High Plains became extinct at the end of the Clovis Period, and sites from the succeeding Folsom Period (11,000 to 10,000 B.P.) indicate a subsistence base focused on the bison (Holliday *et al.* 1983; Hughes and Willey 1978:25; Johnson 1988b). The Folsom occupation in North America was first identified just northwest of the study area at the Folsom site in New Mexico (Figgins 1927), and a number of sites from this period have subsequently been recognized on and adjacent to the Southern High Plains themselves. Like the Folsom "type site," some of the localities producing Folsom-aged material are kill sites, including Lipscomb (Schultz 1943) and Lubbock Lake (Johnson 1988b). Others appear to be campsites, including the Elida site (Hester 1961, 1962; Warnica 1961) and Lake Theo (Harrison and Killen 1978; Harrison and Smith 1975). Unfluted points of the Midland type that seem to be contemporary with fluted Folsom points (Agogino 1969; Rovner and Agogino 1967) were first recognized in the study area at the Sharbauer site (Wendorf *et al.* 1955; Wendorf and Krieger 1959).

Following this period, the cultural chronology is less precise. Projectile points dating to the post-Folsom Paleoindian Period are morphologically quite variable, and type definitions have often been imprecise. One result of this problem has been that poorly defined types have tended to become "catchall" categories, and type names have been somewhat indiscriminately applied to points from widely separated locations on the Great Plains; the Plainview type is a particularly clear example of this (Johnson and Holliday 1980; Knudson 1983). Although there appears to be a relatively distinct sequence of post-Folsom Period point types on the Northwestern Plains (Frison 1978:31–40; Irwin-Williams *et al.* 1973), no such sequence is presently known for the Southern Plains.

Wheat (1972:140–155, 163–164) argues that different types of points are found on the Northern Plains and on the Central and Southern Plains and suggests that the post-Folsom assemblages in the latter region be collectively referred to as the Firstview Complex. This complex includes the San Jon and Firstview types and secondarily includes the Milnesand and Plainview types. An extensive survey of the existing data (Johnson and Holliday 1980) indicates that the distribution of the Plainview type in particular is largely or completely confined to the Southern Plains and that Plainview points date to a relatively short period around 10,000 B.P., although the exact beginning and ending dates of this period are uncertain. San Jon, Firstview, and Milnesand points appear to date after this time (Johnson and Holliday 1980, 1981; Holliday *et al.* 1983, 1985; Wheat 1972).

Plainview sites on the Southern High Plains include the Plainview type site in Plainview, Texas (Guffee 1979; Knudson 1983; Sellards 1952:60–69; Sellards *et al.* 1947), a mass bison kill, Lubbock Lake (Johnson and Holliday 1980;

Holliday and Johnson 1981), and possibly Blackwater Draw (Hester 1972). Marks Beach (Honea 1980), dating to approximately 10,000 B.P. but lacking diagnostic points, may also be a Plainview site. Sites dating after Plainview times fall into the Firstview Period, and include San Jon (Roberts 1942), Milnesand (Sellards 1955; Warnica and Williamson 1968), Rex Rodgers (Hughes and Willey 1978:51–114), Blackwater Draw (Hester 1972), and Lubbock Lake (Bamforth 1985; Johnson and Holliday 1981). Of these four, Milnesand is apparently a mass bison kill, whereas the others are kills of very small numbers of bison. Only the Lake Theo site (Harrison and Killen 1978) has produced evidence of a post-Folsom camp in the study area.

Following this general framework, there are four basic chronological divisions of the Paleoindian Period on the Southern High Plains: Clovis (12,000 to 11,000 B.P.), Folsom (11,000 to 10,000 B.P.), Plainview (ca. 10,000 B.P.). and Firstview (ca. 10,000 to 8000 B.P.). However, for the analysis in the following chapter, it is necessary to combine the Plainview and Firstview periods into a single unit that will be referred to as "Late Paleoindian" because many of the sites recorded in the region can be dated only by the types of projectile points noted by the people who recorded them. Given the lack of precision in the type designations that have been used in the past, it would be potentially misleading to rely heavily on the specific types recorded. The major typological distinctions drawn when these two periods are combined are between the two relatively well-known types of fluted points (Clovis and Folsom) and unfluted Late Paleoindian points. Furthermore, a tripartite distinction between Clovis, Folsom, and Late Paleoindian has been used in several previous studies in the region, and a similar distinction here makes comparisons with this past work more meaningful.

2. CURRENT RECONSTRUCTIONS OF PALEOINDIAN ADAPTATIONS ON THE GREAT PLAINS

The importance of large mammal hunting is stressed in most reconstructions of human adaptations on the Plains during the Paleoindian Period. The great majority of the archaeological sites that date to this period in the region are locations where one or more such mammals were killed and butchered, and the dominance of these sites in the archaeological record has profoundly affected our view of the Paleoindian way of life.

On the Southern Plains and elsewhere during Clovis times, a variety of large extinct fauna occur in clear association with cultural material, including mammoth, horse, camel, and bison (Frison 1978:85–112; Frison *et al.* 1977; Hester 1972; Johnson 1977; Sellards 1952). The Clovis Period is marked as much by the presence of such fauna, particularly mammoths, as it is by fluted points. Later Paleoindian sites are dominated by bison, and kill sites from these periods contain anywhere from one or two to several hundred animals (Agenbroad 1978; Figgins 1927; Frison 1974, 1978:149–191, 1984; Frison and Stanford 1982; Fulgham and Stanford 1982; Hester 1972; Hughes and Willey 1978:51–114; Johnson and Holliday 1980, 1981; Rancier *et al.* 1982; Roberts 1942; Sellards 1955; Sellards *et al.* 1947; Stanford 1978; Wheat 1972).

Several authors have also stressed the importance of a number of other food resources in the Paleoindian diet, noting that the relatively few campsites dating to this period often contain a variety of small and large game (Davis 1953; Forbis and Sperry 1952; Johnson 1977; Wheat 1979; Wilmsen and Roberts 1978). The occasional presence of grinding stones in such sites (Davis 1953; Hester 1962, 1972; Wheat 1979; Wilmsen and Roberts 1978:121–123) further suggests, not too surprisingly, that plants were also eaten. Recognizing these data, several authors have suggested that the heavy emphasis on kill sites in Paleoindian archaeology may have seriously biased our view of the Paleoindian diet and hence of the Paleoindian way of life. Such arguments often cite evidence indicating that the diets of modern hunter–gatherers are usually dominated by plant foods and a wide array of smaller animals (i.e., Greiser 1985).

However, none of the sites producing remains of small animals and evidence of plant exploitation have produced evidence that such foods made up a large proportion of the Paleoindian diet. Merely noting that this diet included a variety of animals and plants without systematically assessing the relative proportions of these foods that were consumed is no less misleading than considering only data from communal kill sites. In fact, there are no currently available data that indicate a level of exploitation of small animals or plants on the Plains that is likely to have produced an amount of food comparable to that obtained from large mammals. For example, Wilmsen and Roberts (1978:47) note a wide range of species represented at Lindenmeier but point out that the data available from the site indicate clearly that bison were the major element in the diet.

The archaeological record for the Paleoindian Period on the Plains, in fact, strongly resembles that from the Late Prehistoric Period (cf. Frison 1967), in that both are dominated by large mammal kill sites and also contain a smaller number of camps producing evidence for a lesser reliance on other foods. There is no doubt that the recent Plains tribes are accurately described as hunting societies, a description that at least partly reflects the paucity of wild plant foods in the grasslands discussed in Chapter 1. Given that the Plains were open grasslands during the Late Prehistoric and Paleoindian periods and that the archaeological records of these two periods are similar, the traditional interpretation of late Pleistocene and early Holocene human groups as primarily hunting peoples seems reasonable.

Several more detailed aspects of the data on Paleoindian adaptations also closely resemble certain aspects of later adaptations. As Chapter 1 argues, the nature of the artifact assemblages found in communal kill sites throughout Plains prehistory indicates a strong continuity in the ways in which human beings prepared and organized themselves for the activities that these sites represent. Similarly, Stanford (1978) has found features within the bone bed at the Late Paleoindian Jones–Miller bison kill in Colorado that closely resemble ritual features known from more recent kills.

Such similarities probably largely reflect the practical requirements of mobilizing a relatively dispersed human population to kill and process large numbers of animals. Other aspects of the archaeological record indicate that there have been important changes in the activity of communal hunting over time. Sites that are comparable to the "classic" Late Prehistoric kills such as Vore (Reher and Frison 1980) or Gull Lake (Kehoe 1973), where a specific location was used for repeated kills over a short period of time, are simply unknown in Paleoindian times. Individual occupations at Paleoindian sites that were used repeatedly for either large or small kills (Frison and Stanford 1982; Hester 1972; Johnson 1988b) are typically separated from one another by gaps several hundred years long. In addition, the products of later kills seem to have been used more completely than those of Paleoindian kills. Skeletons of unbutchered animals and articulated body parts are relatively common in Paleoindian kills, and a number of caches of meat are known that were only partially consumed (Frison and Stanford 1982; Todd 1983).

Recent communal hunts also tended to take cow/calf herds and to occur in the fall, and a major goal of these hunts was to obtain supplies of meat to dry or turn into pemmican for use over the winter (Reher and Frison 1980). In contrast, Frison and Stanford's (1982) analyses of the age structure of animals in Folsom through Late Paleoindian kills at the Agate Basin site in eastern Wyoming indicate that these kills occurred primarily in late fall and winter. They also point out that these kills lack processing areas with extensive evidence for bone grease production, an important step in pemmican manufacture. This suggests that rather than preparing the products of these kills for future use, Paleoindian hunters simply took advantage of the low temperatures during the kill season, froze the partially dismembered carcasses, and lived temporarily in the vicinity of the kill, at least on the Northwestern Plains (Frison 1982, 1984; Frison and Stanford 1982).

Not all Paleoindian kills occurred in winter, though. Summarizing data from post-Folsom Paleoindian kills over a wider region that includes the Northwestern Plains, McCartney (1985) has shown that the communal hunting season appears to extend from early fall into early spring; his data show no evidence for communal kills during the rest of the year. Furthermore, the kills in his sample indicate exploitation of small bull groups as well as larger cow/calf groups.

More comprehensive reconstructions of Paleoindian adaptations than these tend to be based on extremely limited data or simple ethnographic analogy. An overall pattern of seasonal aggregation and dispersion linked to communal hunts is generally inferred or assumed. Wilmsen and Roberts (1978:179), for example, argue that such a pattern accounts for the stylistic variation in projectile points and spatial structure of the Folsom occupations at the Lindenmeier site, and Greiser (1985) predicts that group size should have fluctuated between 25 and 200 people at various seasons throughout the Paleoindian Period (also see Gorman 1972). Other common assertions about this period reflect the dominance of extremely simple hunter–gatherers in the recent ethnographic record, and most scholars assume that Paleoindian social organization was extremely simple, that social group membership was extremely flexible, and that local groups tended to contain about 25 members (Greiser 1985:107; Hester and Grady 1977; Hester 1975).

The range of movement of local groups has also been estimated by considering the sources of the stone tools found in several sites (Hester and Grady 1977; Wilmsen 1973). Reliance on widely separated sources of high-quality raw material has been seen as a Paleoindian adaptation that enhanced mobility by allowing knappers to produce bifaces and other highly standardized tools designed to remain useful over long periods of time, presumably reducing the frequency of visits to quarries (Goodyear 1979).

However, this argument must be clarified somewhat for the Plains data. Although Paleoindian sites throughout the Plains regularly produce tools made from stone that was obtained from quarries a considerable distance away (Bamforth 1985; Frison 1974; Frison and Stanford 1982; Hester 1972; Wheat 1972; Wilmsen 1973), so do sites from later periods. Reher and Frison (1980:127–135) have studied the distance of the sources of the stone used for projectile points in Paleoindian through Late Prehistoric kill sites from the sites in which those points were recovered, and their analysis indicates no notable difference in the size of the area apparently exploited by the earliest and latest hunters at the sites in their sample. Paleoindian groups may have moved around more often than later groups, but Reher and Frison's data suggest that these moves occurred within an area similar to that covered by at least some of the more recent groups.

Few attempts have been made to integrate these relatively limited reconstructions into a more complete picture of Paleoindian ways of life. Greiser's (1985) recent work is one such attempt made for the Central Plains of northeastern and central Colorado, southwestern Wyoming, and western Nebraska and Kansas. Relying primarily on Jochim's (1976) model of hunter–gatherer subsistence and settlement patterns, Greiser attempts to synthesize climatic changes and the available archaeological data for the Paleoindian Period into a single interpretation of late Pleistocene and early Holocene human adaptations.

Briefly, Greiser predicts that increasing seasonality during the Paleoindian Period should have led to an increasing emphasis on stored food to get through the winter, particularly the stored products of large-mammal kills. She also argues that small game and vegetable food should have dominated the diet during spring and summer throughout the period and that large game should have been dominant in fall and winter. Human populations should have been dispersed except when large game was being procured, at which time she expects groups of 100 to 200 people to have formed. She also suggests that larger groups may have persisted through the winter in the later portions of the period as more severe winters restricted human mobility and that settlement locations should have clustered more and more as a deteriorating climate reduced the number of localities that were suitable for occupation. Her analysis of the available archaeological data seems to support these predictions (Greiser 1985:25–35, 54–78, 100–110), but problems with her environmental discussions and certain aspects of her analysis make it difficult to accept her arguments.

Much of her environmental analysis is extremely general and particularly lacks the data needed to support the explicit inferences about seasonal resource availability that Jochim's model demands. For example, she states (Greiser 1985:25, 28) that during the Clovis Period "herd size and composition varied seasonally in accordance with mating habits" and that "the degree of spatial clustering [of large ungulates] was seasonally dictated" but does not state how herd sizes and composition might have varied or how seasonal changes might be related to the spatial distribution of these herds. Several of her specific assertions about the characteristics of subsistence resources on the Plains are also incorrect: bison aggregate for the rut in the summer, not in the late fall as she states, and ungulates do not aggregate in response to deteriorating forage conditions; they disperse (see Greiser 1985:15, 30). In other cases, her reconstructions are either arguably incorrect, as when she assumes that bison herds were as large during the late Pleistocene as in recent times, or are overgeneralizations from limited information, as when she implicitly assumes that all large ungulates divide themselves into bull and cow/calf groups (Greiser 1985:27, 30).

Important aspects of her analysis of the archaeological data are also problematic. Although she states (Greiser 1985:34) that the "estimated proportions" of large game, small game, and vegetable foods in the diet changed from Clovis to later times, she presents neither these estimates nor the data from which they might have been derived. Her inference of increasing reliance on food storage over time on the basis of an increase in the number of bison in single kills (Greiser 1985:106–107) is also difficult to accept. A communal drive kills most or all of the animals in a given herd regardless of the exact number of animals the hunters wanted. If the sizes of individual herds of bison were increasing during the early Holocene, communal hunts would necessarily have taken progressively more animals over time whether or not the hunters were storing or even just eating more meat. Finally, her inferences of the seasons during which a number of Folsom and later campsites were occupied appear to be based directly on the assumptions of her theory (Greiser 1985:99, 105) and therefore cannot provide a test of that theory.

Kelly and Todd (1988) have recently presented another general view of late

Pleistocene (Clovis and Folsom) adaptations in North America. These authors note and attempt to explain several unusual features of the archaeological record from this period that suggest important differences between it and more recent periods. These features include the homogeneity of Clovis assemblages across most of North America, the extensive Paleoindian reliance on high-quality raw materials for tool manufacture and frequent production of bifaces, the frequent direct association of camp- and kill sites and lack of substantial variation in the assemblages found in campsites, and evidence from recent analyses of bison kill sites that much of the meat available from the kills was not used.

They argue that this pattern represents the response of groups of people who had recently entered North America and had not acquired a detailed knowledge of local environments. Lacking this knowledge, these groups could not predict accurately where food and other resources might be found. Furthermore, they argue that the progressive change in the Paleoindian environment would have made it extremely difficult to acquire this knowledge. Kelly and Todd suggest that the response to this problem was to make a kill, camp beside it, and immediately begin searching for another herd of animals. When such a herd was located, the group would move to it, abandoning unused portions of the previous kill. When environmental fluctuations in a local region depleted the game in that region, the hunters simply moved to new lands. In this context, reliance on high-quality raw material and tools such as bifaces that are useful for many tasks and can be used for long periods of time is important, because it is difficult to predict when a group will be near a quarry. They further suggest that some aspects of this theory may apply to post-Folsom adaptations on the Plains.

Kelly and Todd raise a number of important points, but some aspects of their argument are problematic. First, their reconstruction assumes that hunting groups constantly moved into new, unknown territory and away from familiar areas whose productivity had been reduced by temporary environmental fluctuations. However, unless the advancing human population moved as a front, leaving no one behind, such a pattern could have occurred only at the very edges of the occupied area. Away from these edges, human movements would have occurred within known territory. Treating Folsom as well as Clovis groups as pioneers is also questionable: data from South America indicate that human beings penetrated the entire New World by 11,700 B.P. (Krieger 1964:67), implying that the pioneering occupation of North America ended prior to that time and that human beings had been in residence there for several hundred years before the Folsom Period began. It seems extremely conservative to assume that mobile people who depended on hunting for their survival would have taken more than a generation or two to "settle into" their new environments.

Constant moves from kill to kill also imply year-round hunting, which the data cited before show no evidence for. Such moves also suggest that local groups would have constantly maintained the labor force needed to carry out such kills, ruling out a seasonal pattern of aggregation and dispersion. Unpredictable moves through unknown territory would also have made such a pattern extremely difficult to maintain. Contradicting this, Wilmsen and Roberts's (1978) analysis of the data from Lindenmeier appears to indicate that a regular pattern of seasonal aggregation and dispersion existed.

The implicit assumption that group sizes did not fluctuate also leads Kelly and Todd to suggest that Bull Brook, a large fluted-point site in Massachusetts (Byers 1954), was a location that was occupied repeatedly over a longer period of time than the smaller sites around it, which were presumably used only once. However, this interpretation of Bull Brook neglects the site's internal spatial organization. The Bull Brook site consists of a circular arrangement of fairly evenly spaced, discrete concentrations of cultural material, a pattern that suggests to most researchers that it represents a large, probably seasonal, aggregation of distinct social groups (e.g., Gramly 1982:72–78; Grimes *et al.* 1984). If local population size is linked at all to seasonal resource availability, some degree of aggregation and dispersion must be built into Kelly and Todd's hypothesis.

3. SUMMARY

Although neither of these reconstructions seems to be completely acceptable, they both present useful observations about the general Paleoindian adaptation to the Great Plains. The lack of repeated short-term reuse of specific kill sites particularly suggests relatively high mobility, although the repetitive occupation of sites such as Lindenmeier implies that this mobility was tied to specific points on the landscape. However, the relatively well-defined hunting season indicated by the data from communal kills indicates that the need for large groups of hunters was restricted to about half of the year, and the evidence that Paleoindian hunters did not attempt to dry or otherwise preserve large amounts of meat for consumption in later seasons suggests that the food to support large groups may not have been available during the off-season. A degree of seasonal aggregation and dispersion is therefore likely.

It is also likely that Paleoindian population densities on the Plains were substantially lower than those attained by recent societies in the region. Chapter 1 discusses the relative scarcity of plant foods on the Plains and the general implications of this scarcity for human subsistence. Because there were no agricultural societies on the Plains margins from whom Paleoindian hunters could obtain such food, the limit on human population densities must have been lower than in later periods. High mobility and exploitation of large territories may have been linked to this limit as well as to the need to search out game.

Wobst (1974) argues that a minimum population is needed to provide mates for the eligible members of any society. At extremely low population densities, these people will often be spread over a much larger area than would be the case at higher densities, and high Paleoindian mobility may have been partly the result of the need to maintain social contacts with a large number of widely separated social groups. The evidence for unused meat in kill sites and caches may simply indicate that the kills these groups made produced more meat than they could use.

Whether or not this was the case, a pattern of seasonal aggregation and dispersion linked to communal hunts implies that the hypothesis proposed in Chapter 2 linking ungulate ecology and human adaptations should predict important aspects of variation in the Paleoindian way of life. The following chapter derives such predictions from the pattern of environmental change outlined in this chapter and tests them insofar as it is possible on the basis of the available archaeological data from the Southern High Plains and from the Plains in general.

Chapter 11

Paleoindian Responses to Environmental Change on the Southern High Plains

The theory presented in Chapter 2 relating human land use and organization to ungulate herding and migration patterns is clearly relevant to the basic adaptation outlined in the preceeding chapter. However, the nature of the data available on the Paleoindian occupants of the study area puts very different constraints on this section of the discussion than did the data available on recent Plains societies. The relationship proposed here between a region's environment and the organization of the people living in that environment has two distinct components: first, it predicts the distribution of the human population across the landscape and, second, it predicts the effects of this distribution on human organization. In an archaeological context, the first of these predictions is much more easily examined than the second. Reconstructions of prehistoric land use and patterns of aggregation and dispersion are relatively common in archaeological research, but the material correlates of organizational differences among relatively simple societies are rarely studied.

This discussion therefore forms a sort of mirror image to the discussion of the more recent Plains groups. Although the ethnographic record lacks detail on subsistence–settlement patterns but yields useful information on organizational complexity, the archaeological record tends to provide information on the former topic but not on the latter. The focus of this chapter is thus primarily on assessing the changes expected in Paleoindian land-use patterns in response to environmental change and only secondarily on the implications of these changes for the organization of Paleoindian society; it presents these implications mainly as topics for future research.

1. PREDICTING PALEOINDIAN ADAPTATIONS ON THE SOUTHERN HIGH PLAINS

Chapter 10 suggests a basic outline of the Paleoindian way of life on the Plains. Paleoindian societies relied primarily on large game for their subsistence, although plants and smaller animals made up some proportion of their diet. Individual groups of hunters appear to have been extremely mobile; whether or not they directly procured the stone for their tools, they at least had access to lithic resources over large areas. Communal hunts seem to have occurred from fall through early spring. This last point suggests that human aggregations for communal hunts were less seasonal than they were in recent times and that the group required for a successful communal hunt either stayed together for more of the year or dispersed and reformed repeatedly during the hunting season. It is also possible that the size of these groups may have increased over time as the hunters needed to control progressively larger and larger herds of bison.

Although the earliest inhabitants of the Southern High Plains were presumably pioneers penetrating unknown territory, it is not likely that they would have taken long to learn how to make a living in the region. The Paleoindian Period was marked by continuous environmental change, as Chapter 9 discusses. but this change occurred over several thousand years, and the degree of change within the lifetime of any individual human being was probably small. Furthermore, environmental change after the late Pleistocene extinctions had occurred affected a single species—the bison—by altering existing aspects of its behavior rather than by selecting for fundamentally different adaptations.

Human beings can similarly adapt to such a pattern of change by progressively altering their existing ways of life to accomodate progressive changes in their effective environment. The archaeological evidence indicates that Paleoindian hunters relied on the communal hunt as a fundamental means of supplying many material needs throughout the early Holocene. Considering the requirements of this activity outlined in Chapter 1, it is possible to predict how the basic adaptation outlined before should have been articulated with the specific structure of the late Pleistocene and early Holocene environment of the Southern High Plains.

At the end of the Pleistocene, ungulates were abundant, dispersed across the landscape, and tended to move in fairly regular patterns within smaller areas than they did in later times. In response to this, the discussion in Chapter 2 implies that Clovis populations should also have been dispersed. Given the predictable nature of the resource base at this time, it is also conceivable that this population might have formed large groups that relied on small hunting parties for their food. However, a human group that remains in one place for any length of time tends to drive the nearby animals away, which would have continuously increased the effort needed to make a successful kill as the local fauna was hunted out or frightened off. Dispersion in small groups that moved often would eliminate this problem and is the pattern expected here.

After this time, the available animals progressively became fewer and concentrated in larger herds or more patchily distributed. Such a trend should have led the hunters preying on them to form larger groups to provide the labor needed to hunt such herds and to take advantage of the concentrated resource they provided. Given the ease of predicting herd sizes and locations early in this period, aggregations for such hunts should have been relatively regular. Predicting locations for such aggregations in advance should have been reliable, and hunting groups could therefore have relied on a restricted number of prime locations.

These aggregations were probably not very large or of long duration. Pedestrian bison hunters inevitably lose some of the herds they try to entice into their traps, and if the bison live in smaller groups, more such groups must be trapped to get large amounts of meat. Contacting more herds would have frightened more of them away, eventually driving most or all of the available animals away from the aggregation site.

As the bison herds became progressively larger, the productivity of any single mass kill would have increased. At the same time, these herds became more mobile and less predictable from year to year, making it more difficult to locate them at any given time and to predict in advance where they might be found. An increasingly patchy resource distribution should have led to larger human aggregations, but choosing locations for these aggregations must have been more difficult, requiring more detailed local environmental information and probably leading to a more opportunistic use of a greater variety of locations. Furthermore, as environmental conditions fluctuated more from year to year there must have been increasing numbers of years when conditions could not support any large aggregations. However, when aggregations were held, they may have been larger and persisted longer because of the greater volume of meat and hides produced in a single drive.

These trends also have predictable effects on social heterogeneity, although, as is noted before, these predictions are largely untestable at present. The expected dispersion of the Clovis Period population should have inhibited complexity. The high predictability of faunal resources during this period should have supported relatively regular associations of people, but the expected small size of social groups should not have required extensive formal mechanisms of social control. Clovis Period heterogeneity should thus have been relatively low. Resource predictability remained relatively high during the Folsom Period, and the success of the seasonal communal hunts must have required relatively disciplined hunters. The increased patchiness of the resource base at this time should have also increased local group size substantially, and these two trends together should have increased heterogeneity substantially above the levels existing in the preceding period.

Early Holocene climatic changes should have disrupted this trend. The size of seasonal aggregations should have grown during this period, but the regularity of such aggregations must have decreased. Annual fluctuations in environmental conditions would have tended to inhibit aggregation in many years, and decisions to aggregate should have been made over the short term rather than months in advance. Regular associations of specific individuals would have been difficult to maintain under these conditions, and heterogeneity should thus have decreased. This trend would have made the larger aggregations of people expected for this period relatively unstable, and disputes and other problems must have inhibited the ability of such groups to stay together for very long.

These trends have several archaeological implications. First, larger social groups staying in one place longer should generally produce larger sites with denser and more diverse artifact assemblages than smaller groups or shorter occupations, particularly if a single location is used repeatedly. Larger, denser, more diverse sites should thus be more numerous when aggregations are longer or last longer. Second, aggregations planned long in advance are likely to be held at the "best" locations in a region, taking *best* to mean locations where the likelihood of a successful hunt is high and where the other resources needed to support an aggregation (such as wood for fires or construction) are available. On the Southern High Plains, areas near water sources, particularly permanent water sources, provide locations such as these and should have been used preferentially when environmental conditions allowed. Aggregations planned in the short term because of more variable environmental conditions should rely more on immediate knowledge of local conditions and are therefore likely to occur at a wider variety of locations. In addition, deteriorating forage conditions should have forced ungulates and therefore the humans preying on them toward the wetter eastern portions of the study area over time and should also have increased the importance of areas around reliable water supplies.

2. THE ARCHAEOLOGICAL EVIDENCE

Data on which these predictions can be assessed are available from the published record of research on the Paleoindian occupation of the study area in particular and the Great Plains in general and from site data filed in archives on the Southern High Plains. These data are not currently capable of rigorously testing these predictions in detail, but it is possible to see if the available data and the predictions are consistent with one another.

Published and archival information have very different limitations. First, the majority of the recent intensive work on the Paleoindian Period on the Plains has been conducted outside of the present study area. Appealing to this work to assess hypotheses proposed specifically for the Southern High Plains assumes that similar adaptive changes occurred throughout the entire Plains region. The uniformity of early Holocene changes in bison morphology throughout the Plains suggests that changes in the effective environment of the whole region were similar, lending some support to this assumption.

In contrast, the archival site data available here are directly relevant to the study area but are less completely documented than the sites described in the published record. These data include all sites producing Paleoindian projectile points that were on file as of 1981 at Texas Technological University, in Lubbock, Texas, West Texas State University, in Canyon, Texas, and the Bureau of Land Management (BLM), in Roswell, New Mexico. These sites were recorded over approximately 40 years by amateurs and professionals with a wide variety of research objectives, and the information available on specific sites ranges from simple notes that points of a particular kind were found in a general area to relatively detailed descriptions of site locations and contents that can be supplemented by published reports. Such diverse data clearly cannot support detailed or complex analysis, but using a simple site classification and quantitative methods that make minimal assumptions about the structure of the available data, these records can provide useful information.

This analysis classifies the sites into four categories: multiple-activity sites, limited activity sites, kill sites, and isolated artifacts. Multiple-activity sites tend to be large and to contain dense and diverse concentrations of cultural materials. In contrast, limited activity sites are usually small and/or light scatters of flakes with few or no other types of artifacts present. Multiple activity sites appear to represent either larger groups, longer occupations, or more frequent reoccupation than limited activity sites. Information on site sizes was often not available, and most determinations were made on the basis of site contents. The distinction between the relatively dense, diverse contents of multiple activity sites and the sparser, more restricted contents of limited activity sites is relatively easy to draw. A comparison of such distinctions made independently for a number of sites by two individuals revealed no disagreements.

Kill sites are a specific class of limited activity site distinguished by abundant animal remains associated with few tools other than projectile points. Kill sites whose faunal contents have been destroyed by postdepositional processes would be classified as limited activity sites because of the relatively small number and restricted range of flaked stone artifacts that such sites typically produce. The class of isolated artifacts is self-explanatory. The utility of a classification recognizing a major distinction between multiple and limited activity sites made on the basis of site size and contents has previously been shown by Read (1983) in an analysis of archival data and Kvamme (1984) in an analysis of data collected during a systematic field survey.

All sites were classified by a single individual to maintain consistency. Each site was assigned to a time period on the basis of the type(s) of points identified by the individual(s) who recorded it. Because of the potential typological problems noted in Chapter 11, only three periods are recognized for these data: Clovis, identified by the presence of Clovis points, Folsom, identified by the presence of Folsom or Midland points, and Late Paleoindian, identified by the presence of any other type of Paleoindian point. If points dating to more than one period were noted for a single site, the site class indicated by the full range of other material present was recorded for each period. This last procedure neglects the possibility that a site's function changed over time (cf. Binford 1982), but the presently available data simply present no way to address this problem.

When the location of a site could be identified accurately, two additional variables were coded, one describing its topographic setting and one describing the nature of the water source closest to it. In general, to code these variables, the sites were plotted on USGS 7.5- or 15-minute topographic maps. For some sites that could not be so plotted, verbal descriptions on the site records provided a basis for inferring one or both of these variables. Ambiguity about the locations of a number of sites causes the total numbers of sites from the various periods to vary somewhat from table to table.

The first variable notes the general topographic setting in which the site was found, distinguishing between sites located on the open plains and those immediately adjacent to rivers, other drainages, small deflation basins, and lakes. The general lack of topographic diversity in most, although not all, of the study area combined with the relatively small sample of sites suggests that this simple classification of site surroundings is adequate for this analysis.

The second variable recorded is the nature of the closest type of water source to the site. This variable distinguished between rivers, currently permanent streams, currently intermittent streams, springs, small playas, and lakes. These codes are clearly similar to those for the preceding variable, but they add additional information for those sites that are not immediately adjacent to the water source that is closest to them.

3. PALEOINDIAN SITE TYPES AND DISTRIBUTIONS ON THE SOUTHERN HIGH PLAINS

The data available here include information on a total of 90 sites, with 19 dated to the Clovis Period, 26 dated to the Folsom Period, and 46 dated to the Late Paleoindian Period. Because apparently isolated points may truly be isolates or may be the only portion of a partially buried site noted by the individuals who recorded them (cf. Gramly 1982:79), they play no part in this analysis. This reduces the total sample of sites to 83.

Table 11-1 presents the frequencies of multiple activity, limited activity, and

Table 11-1. Frequencies of Sites by Type and Period

Period	Site type			Total
	Multiple activity	Limited activity	Kill	
Clovis	0[a]	11	6	17
	(2.9)[b]	(10.8)	(3.3)	
Folsom	9	13	3	25
	(4.2)	(16.0)	(4.8)	
Late Paleoindian	5	29	7	41
	(6.9)	(26.2)	(7.9)	
Total	14	53	16	83

[a]Observed frequency.
[b]Expected frequency.

kill sites by cultural period, and these frequencies depart from random expectation ($\chi^2 = 12.6$, $df = 4$, $.02 > p > .01$). The major deviations from expectation are in the total absence of multiple activity sites and excess of kill sites in Clovis times and the excess of multiple activity sites in Folsom times. There is also a decrease in the proportion of multiple activity sites in the sample from 36% in Folsom times to 12% in Late Paleoindian times.

The available data also suggest several shifts in the locations of these sites over time. Tables 11-2 through 11-7 show the frequencies of the three types of sites defined here by topographic setting and the nature of the nearest water source for each period. Although these frequencies are clearly too low for statistical analysis, it is possible to collapse these tables and observe several interesting patterns.

The topographic data can be collapsed to distinguish between sites located directly on water-related features and sites located on the open plains. Water sources are good places to find game and often provide the topography needed to trap them. They also offer other resources, such as shelter from the elements and wood for fuel and equipment manufacture, which are often important determi-

Table 11-2. Frequencies of Clovis Sites by Type and Topographic Setting

Type	Setting				Total
	Open plains	Playa	Lake	Stream	
Limited activity	8	1	1	1	11
Kill	1	1	0	4	6
Total	9	2	1	5	17

Table 11-3. Frequencies of Clovis Sites by Type and the Nature of the Closest Water Source

Type	Water source: Permanent stream	Intermittent stream	Spring	Playa	Total
Limited activity	1	3	0	7	11
Kill	0	4	0	2	6
Total	1	7	0	9	17

nants of hunter–gatherer site locations (cf. Jochim 1976:47–51). Water sources should therefore be preferred locations for sites, as was noted earlier.

The data on the nearest water source can be considered in two ways. First, a basic distinction can be drawn between small deflation basins and other types of sources. The basins are fed solely by rainfall and are thus strictly seasonal resources; other sources are less closely linked to precipitation. Basins should thus have been important to human occupation during only part of the year, and the proportion of the year during which they were important should have decreased over time as precipitation became more seasonal.

Second, a distinction can be drawn between perennial sources (lakes, permanent streams, rivers, and springs) and intermittent sources (intermittent streams and deflation basins). Although this distinction is based on modern conditions, only the difference between permanent and intermittent streams is likely to have changed over time: rivers, lakes, and springs were presumably permanent water sources in the past, and deflation basins, as was just noted, were not. Although geologic evidence (see Chapter 10) indicates that increased surface water during the Paleoindian Period created through-flowing streams at times in drainages that are now intermittent, the *relative* difference in the reliability and volume of the water, and thus in many of the resources associated with them

Table 11-4. Frequencies of Folsom Sites by Type and Topographic Setting

Type	Setting: Open plains	Lake	Stream	Total
Multiple activity	4	3	2	9
Limited activity	9	0	3	12
Kill	1	0	2	3
Total	14	3	7	24

Table 11-5. Frequencies of Folsom Sites by Type and the Nature of the Closest Water Source

	Water source					
Type	Permanent stream	Intermittent stream	Spring	Playa	Lake	Total
Multiple activity	1	0	1	1	2	5
Limited activity	0	5	0	6	0	11
Kill	0	3	0	0	0	3
Total	1	8	1	7	2	18

(such as wood and vegetable foods) in these drainages, indicated by the terms *permanent* and *intermittent* is likely to have been constant over time. In general, permanent water sources should have been preferred over intermittent sources for occupation, particularly occupation by larger numbers of people.

Tables 11-8 through 11-10 show the frequencies of Clovis Period sites collapsed in these three ways. None of these tables shows any notable deviations from random expectation (Table 11-8: $\chi^2 = 3.0$, $df = 1$, $.1 > p > .05$; Table 11-9: $\chi^2 = 0.51$, $df = 1$, $.5 > p > .25$; Table 11-10: $\chi^2 = 0.71$, $df = 1$, $.5 > p > .25$; all three chi-square values were computed using Yates's correction). Considering the small sample of sites available, the moderately low probability for Table 11-8 may suggest that kill sites are associated with water sources and that other sites occur on the open plains (cf. Cowgill 1977), but this pattern is as likely to reflect the vagaries of site preservation as the patterned behavior of past human beings, because faunal remains are obviously more likely to be buried and thus preserved near water than on the uplands.

The Folsom Period sites show several patterns similar to those just noted, with the frequencies in most tables staying close to random expectation (Table 11-11: $\chi^2 = 2.11$, $df = 1$, $.25 > p > .1$; Table 11-12: $\chi^2 = 0.48$, $df = 1$, $.5 > p > .25$; Table 11-12 uses Yates's correction). Kill sites are not included in these

Table 11-6. Frequencies of Late Paleoindian Sites by Type and Topographic Setting

	Setting				
Type	Open plains	Playa	Lake	Stream	Total
Multiple activity	2	0	1	1	4
Limited activity	22	1	3	3	29
Kill	1	0	0	6	7
Total	25	1	4	10	40

Table 11-7. Frequencies of Late Paleoindian Sites by Type and the Nature of the Nearest Water Source

	Water source				
Type	Permanent stream	Intermittent stream	Playa	Lake	Total
Multiple activity	0	3	1	1	5
Limited activity	1	8	14	2	29
Kill	0	6	0	1	7
Total	1	17	15	4	41

tables because they are too few for analysis. Despite the lack of association between multiple or limited activity sites and specific types of water sources (Table 11-12), Table 11-13 indicates that there are differences in the proportions of these sites near permanent and intermittent sources ($\chi^2 = 7.4$, $df = 1$, $.01 > p > .005$ using Yates's correction). These data indicate that, during Folsom times. multiple activity sites tended to be located near permanent water and limited activity sites tended to be located near intermittent water.

The Late Paleoindian data show a similar but stronger association between kill sites and water sources (Table 11-14: $\chi^2 = 9.6$, $df = 2$, $.01 > p > .005$) to that suggested by the Clovis Period data. In contrast to the Clovis data, which otherwise showed a random pattern of site locations, the Late Paleoindian kill sites also tend to be associated with sources of water other than deflation basins, and limited activity sites may be located more frequently near such basins (Table 11-15: $\chi^2 = 8.02$, $df = 2$, $.25 > p > .01$). Finally, in contrast to the Folsom data, there is no association of multiple or limited activity sites with permanent or intermittent water sources (Table 11-16: $\chi^2 = 0.2$, $df = 2$, $.95 > p > .9$).

There is also evidence that the distribution of sites within at least part of the

Table 11-8. Frequencies of Clovis Sites by Type and Location on a Water-Related Topographic Feature or the Open Plains

	Setting		
Type	Open plains	Other	Total
Limited activity	8[a] (5.8)[b]	3 (5.2)	11
Kill	1 (3.2)	5 (2.8)	6
Total	9	8	17

[a]Observed frequency.
[b]Expected frequency.

Table 11-9. Clovis Sites by Type and Association with Small Playas and Other Water Sources

Type	Water source: Playa	Water source: Other	Total
Limited activity	7[a] (5.8)[b]	4 (5.2)	11
Kill	2 (3.2)	4 (2.8)	6
Total	9	8	17

[a]Observed frequency.
[b]Expected frequency.

study area changed from period to period. Hester (1975; also see Wendorf and Hester 1962) gathered information from local amateur archaeologists on the locations and contents of Paleoindian sites on the central portion of the Llano Estacado and field-checked a small number of these sites. This work identified a total of 80 sites in Texas and New Mexico, which were classified as camps, kill sites, and isolated artifacts, and it divided these sites chronologically on the same basis as that which is used for the sample of sites analyzed here. The camp/kill distinction was drawn on essentially the same grounds as the distinction drawn here between kill sites and multiple or limited activity sites; Hester's "camp" category therefore combines these two later classes. The New Mexico sites are recorded in the BLM files and are therefore available here, but the Texas sites are not filed anywhere in Texas and are therefore not in the present sample.

Hester and Grady (1977) have considered the general spatial distribution of these sites, particularly emphasizing the degree to which the sites are clustered within their study region, which includes much of the Llano Estacado subsection

Table 11-10. Clovis Sites by Type and Association with Permanent and Intermittent Water Sources

Type	Water source: Permanent	Water source: Intermittent	Total
Limited activity	1[a] (0.7)[b]	10 (10.3)	11
Kill	0 (0.3)	6 (5.7)	6
Total	1	16	17

[a]Observed frequency.
[b]Expected frequency.

Table 11-11. Folsom Period Multiple and Limited Activity Sites by Type and Location on a Water-Related Topographic Feature or the Open Plains

Type	Setting: Open plains	Setting: Other	Total
Multiple activity	4[a]	5	9
	(5.6)[b]	(3.4)	
Limited activity	9	3	12
	(7.4)	(4.6)	
Total	13	8	21

[a]Observed frequency.
[b]Expected frequency.

of the present study area (Figure 11-1). To measure the degree of site clustering, they applied a nearest neighbor statistic (*Rn*) equal to:

$$2d\sqrt{n/a}$$

where d is the mean distance between sites and their nearest neighbors, a is the size of the area within which those sites are contained, and n is the number of sites analyzed. This statistic ranges between 0.0 and 2.15, with a score of 0.0 indicating absolute identity between the locations of all sites, a score of 1.0 indicating a random distribution, and a score of 2.15 indicating a perfectly even distribution of sites. Table 11-17 presents the results of Hester and Grady's analysis by site type for each of the periods they recognize.

The trend in campsites is from a nearly random Clovis distribution to a progressively more clustered pattern in the Folsom and Late Paleoindian periods.

Table 11-12. Folsom Period Multiple and Limited Activity Sites by Type and Association with Small Playas and Other Water Sources

Type	Water source: Playa	Water source: Other	Total
Multiple activity	4[a]	1	5
	(3.1)[b]	(1.9)	
Limited activity	6	5	11
	(6.9)	(4.1)	
Total	10	6	16

[a]Observed frequency.
[b]Expected frequency.

Table 11-13. Folsom Period Multiple and Limited Activity Sites by Type and Association with Permanent and Intermittent Water Sources

Type	Water source		Total
	Permanent	Intermittent	
Multiple activity	4[a]	1	5
	(1.3)[b]	(3.7)	
Limited activity	0	11	11
	(2.8)	(8.2)	
Total	4	12	16

[a]Observed frequency.
[b]Expected frequency.

Kill sites, in contrast, shift from relatively clustered to nearly randomly distributed. Although the total number of sites in the sample increases over time (Table 11-17), possibly because younger sites are more likely to be found than older sites, the statistic used here takes the number of sites in the sample into account. Later sites therefore do not appear more clustered simply because more of them are packed into a fixed area. Considering Hester and Grady's (1977:84–86) maps of the distributions of sites in their sample suggests that much of the increased clustering of the campsites is due to more intensive occupation of two regions, one around a group of pluvial lakes and the other just below the caprock escarpment southeast of Lubbock.

To summarize, the data on site types and distributions in the study area show an absence of multiple activity sites in the Clovis Period and a greater

Table 11-14. Late Paleoindian Sites by Type and Location on a Water-Related Topographic Feature or the Open Plains

Type	Setting		Total
	Open plains	Other	
Multiple activity	2[a]	2	4
	(2.5)[b]	(1.5)	
Limited activity	22	7	29
	(18.1)	(10.9)	
Kill	1	6	7
	(4.4)	(2.6)	
Total	25	15	40

[a]Observed frequency.
[b]Expected frequency.

Table 11-15. Late Paleoindian Sites by Type and Association with Playas and Other Water Sources

Type	Water sources		Total
	Playa	Other	
Multiple activity	1[a]	4	5
	(2.0)[b]	(3.0)	
Limited activity	14	11	25
	(10.2)	(14.8)	
Kill	0	7	7
	(2.8)	(4.2)	
Total	15	22	37

[a]Observed frequency.
[b]Expected frequency.

proportion of such sites in the Folsom Period than is found in the Late Paleoindian Period. The Folsom data also indicate a relatively strong association between multiple activity sites and permanent water sources and limited activity sites and intermittent water sources. This association is absent in the Late Paleoindian Period, but the data from this period suggest that Late Paleoindian kill sites are associated with streams rather than deflation basins and that limited activity sites tend to be found at these basins. Finally, kill sites become progressively more dispersed, and other sites become progressively more clustered throughout the Paleoindian Period.

These patterns generally fit very well with the predictions outlined before. The absence of multiple activity sites in the Clovis Period is consistent with the expected higher degree of population dispersion at this time than in later periods.

Table 11-16. Late Paleoindian Sites by Type and Association with Permanent and Intermittent Water Sources

Type	Water source		Total
	Permanent	Intermittent	
Multiple activity	1[a]	4	5
	(0.7)[b]	(4.3)	
Limited activity	3	22	25
	(3.4)	(21.6)	
Kill	1	6	7
	(0.9)	(6.1)	
Total	5	32	37

[a]Observed frequency.
[b]Expected frequency.

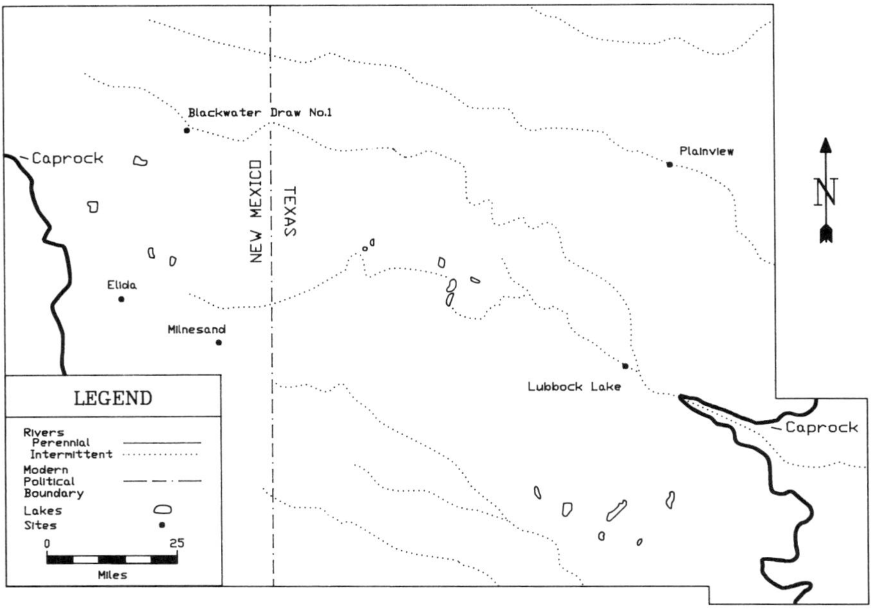

Figure 11-1. Portion of the study area examined by Hester and Grady (1977).

Haynes (1980:118–119) has similarly noted that Clovis campsites appear to be smaller than those dated to later portions of the Paleoindian Period. The meaning of the decrease in the proportion of multiple activity sites from Folsom to Late Paleoindian times is ambiguous. This pattern could be caused by a shift back toward more Clovislike conditions of population dispersal, as is predicted here, but it could also result from more extensive reuse of a relatively few favored

Table 11-17. Degree of Clustering of Paleoindian Sites in the Central Llano Estacado by Period and Type[a]

	Site type			
	Camp		Kill	
Period	Number of sites	*Rn*	Number of sites	*Rn*
Clovis	8	0.89	5	0.46
Folsom	15	0.63	7	0.62
Late Paleoindian	18	0.59	12	1.17

[a]From Hester and Grady 1977:82–83.

locales for occupation—the opposite of the pattern predicted for the Late Paleoindian Period. Clearly, this problem can be resolved only by obtaining more detailed information on these sites than is currently available.

The changes in the pattern of site location also conform to the predicted trends and may shed some light on this issue. The association of multiple activity sites with permanent water sources and limited activity sites with intermittent water sources during Folsom times is consistent with a pattern of preferentially aggregating in particularly productive locations and using less desirable locations for smaller camps or for other purposes. The increased mobility and less predictable migration patterns of bison in the Late Paleoindian Period would have forced hunters more and more to take herds where they could find them, and the absence of an association between multiple and limited activity sites and different types of water sources in Late Paleoindian times may reflect the more flexible settlement pattern expected under such conditions. Such a pattern conflicts with the possibility noted before that the Late Paleoindian decrease in the proportion of multiple activity sites in the sample results from more frequent reuse of preferred locations, suggesting instead that it may represent an increase in the overall degree of population dispersion in the region.

It is also possible that the relative absence of Late Paleoindian kill sites near the small upland deflation basins reflects the shorter season during which these locations would have been potential watering sources for the bison. The small basins in the region would have contained water mainly during the summer, and Paleoindian communal hunts appear to have been held from fall into late winter, as was discussed in Chapter 10. However, it is also possible that there are differences in sedimentation rates in basins and stream channels and that bones left near basins may be exposed on the surface longer, increasing the chances that they will be destroyed rather than buried.

The increased clustering of sites other than kill sites indicated by Hester and Grady's analysis is predicted directly by Horn's model (discussed in Chapter 2) as a response to a more mobile and aggregated resource base. The areas in which sites tend to concentrate over time are particularly suited to communal hunting: the lakes especially were permanent sources of water that would have attracted and concentrated herds of bison. The area below the caprock was probably also better watered by runoff from the Llano Estacado and the springs along the escarpment than the uplands. This latter area also provides many canyons and other locations where herds can be trapped (cf. Frison 1978:150, 193, 202; Hughes and Willey 1978:63–64). An increased emphasis on occupation of these areas was predicted before as a response to increasing dessication of the more western parts of the study area and to the increased use by the bison of permanent water sources as the upland deflation basins became less reliable. The increasingly random distribution of kill sites in the region may also reflect the increasingly unpredictable movements of the herds.

4. OTHER DATA

Data from the Plains and adjacent areas as a whole also provide some support for the predictions outlined here. First, the expectation that individual bison kills should have become progressively more productive during the early Holocene is supported by Judge's (n.d.) analysis indicating that the mean number of bison in communal kills throughout the Plains increased from 27 in Folsom times to 129 in Late Paleoindian times. Second, the nature of the Lindenmeier site in Colorado (Wilmsen and Roberts 1978) fits very well with the general expectations about the effects of the predictability of the Folsom Period environment on human settlement.

The Lindenmeier site appears to be the remains of a series of relatively brief human occupations that occurred over a fairly short period of time. Analysis of the spatial distribution of material in the site identified a series of discrete concentrations of artifacts and faunal debris that apparently correspond to individual structures. Two major areas of occupation were defined within the site, each containing several of these concentrations. All of the concentrations within each of the two areas produced morphologically identical projectile points, but a number of morphological differences are evident when points from the two areas are compared with one another.

Wilmsen and Roberts (1978:179) interpret these patterns as indicating that the Lindenmeier site was a location where several distinct social groups came together, probably seasonally. The consistent spatial distribution of stylistically different points suggests that the same groups resided in specific portions of the site each time they arrived, implying a very regular association of specific people at each aggregation, although the number of these groups present at the site at any one time is unknown because the site was not completely excavated. Wilmsen and Roberts also suggest that these aggregations may have occurred annually. If this interpretation is correct, it implies that these groups were able to come together and support themselves, apparently by hunting bison, at a specific point on the landscape year after year. Just such a pattern was predicted before for the Folsom occupation of the Southern High Plains, and, if this prediction can be extended to the Lindenmeier region, it helps to account for the pattern seen there.

The predictions about the ways in which Paleoindian land-use and aggregation patterns in the study area and possibly elsewhere should have changed in response to late Pleistocene and early Holocene environmental change are thus consistent with the available data. The predicted effects of these changes on human organization are more difficult to examine. Social heterogeneity cannot be assessed directly on the basis of the existing data, but some tentative evidence about changes in the degree of social control can be inferred from data drawn from Paleoindian and later sites on the Plains and adjacent regions as a whole.

These data relate to the organization of the butchering effort represented in kill sites. In recent communal bison kills, the animals killed by individual hunters were identified, often by distinctive markings on the arrows used to dispatch them, and each hunter or his wife butchered his own kill, although the products of the kill were often distributed to the camp as a whole (Wheat 1972). In contrast, the bone beds in a number of Paleoindian kill sites indicate that animals were butchered cooperatively.

This inference is based on the presence in a number of sites of relatively discrete piles of bones that include similar anatomical parts of several animals. Lorraine (1968:105) refers to the behavior suggested by this pattern as "assembly-line" butchering, and Wheat (1972:107) notes the contrast between this pattern and the individualistic pattern indicated in the historic record. Table 11-18 lists Paleoindian and later communal kill sites where this aspect of the structure of the bone bed is discussed in the published report.

Several problems reduce the number of sites that can be included in this table. First, the data it presents refer to the distribution of bones at the actual scene of the kill, which presumably reflects the pattern of immediate butchering of the animals. Data from sites such as Jurgens (Wheat 1979), to which parts of dismembered bison were taken for more complete processing, are therefore excluded. The Colby mammoth site (Frison 1978:91–110), the Paleoindian levels at the Agate Basin site (Frison and Stanford 1982), and the Carter/Kerr-McGee site (Frison 1984) are also excluded because the structure of the bone beds at these localities appears to represent food storage rather than butchering activities at the scene of the kill.

The available data are further reduced because many reports that present detailed analyses of the characteristics of the bones recovered from a kill fail to

Table 11-18. Evidence for Cooperative Butchering in Paleoindian and Later Kill Sites

Site	Location	Date	Bone stacked?	Reference
Murray Springs	Arizona	11,600 B.P.	No	Hemmings 1970
Miami	Texas	Clovis	No	Sellards 1952
Bonfire Shelter	Texas	10,200 B.P.	Yes	Dibble 1968
Olsen-Chubbock	Colorado	10,200 B.P.	Yes	Wheat 1972
Casper	Wyoming	10,000 B.P.	Yes	Frison 1974
Jones-Miller	Colorado	10,000 B.P.	Yes	Stanford 1978
Hudson-Meng	Nebraska	9,800 B.P.	Yes	Agenbroad 1978
Mona Lisa	Alberta	8,000 B.P.	No	Wilson 1974
Itasca	Minnesota	7,000–8,000 B.P.	No	Shay 1971
Bonfire Shelter	Texas	2,650 B.P.	No	Dibble 1968
Ruby	Wyoming	1,670 B.P.	No	Frison 1971

discuss the spatial distribution of the butchered bone within the site and because early excavations of many important sites provide virtually no data on the bone beds at all. Finally, Saunders (1980) has argued that Clovis elephant kills fall into two groups: "catastrophic" kills of whole family units, represented at Lehner Ranch, Dent, and Miami, and sites where the remains appear to be the accumulation of a series of individual kills. Only Saunders's first category is relevant here because catastrophic kills are more likely to represent a communal effort comparable to that involved in large bison kills than the latter. Of these sites, only the bone bed at Miami was published in sufficient detail for this discussion.

One pattern in these data is clear: the post-Paleoindian sites for which data are available show no evidence for the degree of cooperative effort in butchering indicated at earlier sites such as Olsen–Chubbuck. At the Middle Prehistoric Ruby site in Wyoming, for example, Frison (1971:87) explicitly states that the evidence indicates that "processing was carried out at the family level." Lorraine (1968:103) found a clear decrease in the degree to which specific skeletal elements from different animals occur within single excavation units at Bonfire Shelter in southwest Texas between kills dated to 10,200 B.P. and 2,645 B.P., suggesting a decrease in the level of cooperation in the butchering process.

One major environmental characteristic of the post-Paleoindian period appears to be the persistence of more variable climatic and, presumably, forage conditions from year to year. The discussion in Chapter 2 suggests that this should generally inhibit the development of a high degree of social control by reducing the regularity of large social aggregations. If stacking of skeletal elements within the bone bed does indicate a more cooperative organization of butchering, the data in Table 11-18 indicate that such an organization was confined to the Paleoindian Period. This is consistent with the general prediction that more variable environments, such as those existing after the end of the Paleoindian Period, should be associated with weaker forms of social control.

The pattern within the Paleoindian Period is unfortunately incomplete. The limited Clovis data show no evidence for cooperative butchering as would be expected. However, many Clovis kills are of elephants, and butchering even a single animal the size of an elephant may require cooperative efforts and therefore might not produce discard patterns like those seen in bison kills. The bison bone at Murray Springs (Hemmings 1970) was not stacked, but this bone represents only six individuals. Data on the structure of the bone beds on pure Folsom Period kills do not currently exist, although the bone bed at Bonfire Shelter produced a Folsom point associated with several Plainview points. The dates from later Paleoindian bison kill sites for which good evidence is available cluster around 10,000 B.P., and all of them show evidence for cooperative butchering. The Itasca site in Minnesota dates to the very end of the Paleoindian Period and shows no evidence for cooperative butchering, but Shay (1971:32–37) argues that the distribution of bones in the excavated area results from natural redeposition rather than patterns of human discard. The Mona Lisa site (Wilson 1974) also dates late in the Paleoindian Period and shows a similar pattern, but seems to involve a relatively small number of bison. Most of the sites that might have provided information on the later portions of the Paleoindian Period were excavated before the importance of recording detailed information on the structure of the bone beds was recognized.

Adequate data on which to examine the degree of cooperation in butchering over the whole Paleoindian Period thus do not exist. The available data do suggest that there is a fundamental difference between the organization of butchering in Paleoindian kills dated around 10,000 B.P. and the organization of butchering in Archaic and later kills, and this difference fits with the general expectations of the predictions presented here. Although there may also be a difference between Clovis and the butchering patterns seen in Paleoindian sites dated around 10,000 B.P., the available data are not clear. The actual pattern of change in this activity within the Paleoindian Period cannot be assessed more accurately until more data are available.

5. SUMMARY

This chapter proposed that there should have been a constant increase in aggregation size from Clovis through Late Paleoindian times in response to an increasingly patchy distribution of animals on the Southern High Plains. Aggregations should also have become less predictable over time, particularly after the Folsom Period, as herd movements and thus the communal hunt became less predictable. Social complexity should have increased from Clovis to Folsom times and decreased thereafter.

The available archaeological data from the study area and other parts of the Plains are consistent with these expectations, although they do not provide a rigorous test of them. Multiple activity sites are absent during the Clovis Period and are more frequent during the Folsom Period than the Late Paleoindian Period, probably indicating an increase in the degree of social aggregation at the end of the Pleistocene and a subsequent decrease in the regularity of these aggregations thereafter. A high degree of predictability in the location and composition of these aggregations during Folsom times is also indicated by the nature of the Lindenmeier site in Colorado.

The implications of these changes for human land-use patterns were also supported. Changes in the locations favored for sites and in the overall distribution of sites within at least part of the study area are in accord with the predictions outlined in the first part of the chapter. Evidence for organizational change is extremely limited but does not conflict with the expectations outlined here.

The available data thus suggest strongly that there were important changes in several aspects of human adaptations on the Southern High Plains between 12,000 and 7,000 B.P. Although additional research is needed to substantiate these changes, the analyses presented here provide both a clear direction for such research and a strong justification for conducting it.

Chapter **12**

Summary and Conclusions

This study has examined the relationship between the availability of resources in a region, the ways in which human beings exploit those resources, and the effects of these patterns of exploitation on human organization. It proposed that social structure should be more complex when larger groups of people come together more regularly and stay together longer, and that larger, more regular, and longer human aggregations should occur when resources are abundant and concentrated predictably at a small set of locations within a region. Greater complexity under these circumstances should be required to mediate group decisions and provide more effective social control.

Translated into the key resources available on the Great Plains, this theory predicts that larger, more regular, and longer human aggregations should have been held when herds of large ungulates were larger, less mobile, and more regular in their migration and aggregation patterns. More complex societies should have existed where these conditions obtained. After discussing the factors controlling bison and other ungulate herding and mobility patterns, this study considered the implications of this prediction for variation in recent hunter–gatherer adaptations in different parts of the Plains and for changes in such adaptations over time on the Southern High Plains.

A detailed examination of the nature of climatic and hence forage variation on the recent Plains grasslands indicates that there were probably substantial differences in bison adaptations in different parts of the region. Specifically, herds in the southwest should have been smaller, fewer, more mobile, and less predictable in their movements and seasonal aggregation patterns than those farther to the northeast. Documentary evidence on seventeenth- through nineteenth-century bison adaptations and more detailed information on modern bison support these expectations.

Using a climatic index summarizing the important determinants of variations in bison adaptation, this study then examined the available ethnographic data on selected aspects of social complexity on the recent Plains. This examination indicated that more complex aboriginal societies were found in those portions of the Plains where climatic conditions should have supported larger, less mobile, and more predictable herds of bison, as the theory presented here predicts. This relationship between ecology and organization also illuminates several aspects of the tribal migrations across the Plains between A.D. 1650 and 1850.

Having shown that the predictions of the theory at issue here fit well with the data available on the recent occupants of the Plains, this study then turned to an archaeological example of environmental and cultural change over time, examining Paleoindian adaptations on the Southern High Plains of western Texas and eastern New Mexico between 12,000 and 8,000 B.P. Although data bearing directly on human organization during the Paleoindian Period in this region, or anywhere else in North America, are scarce, published research and archival site data provide a basis for examining the land-use patterns that should have affected social complexity.

Substantial climatic change on the Southern High Plains at the end of the Pleistocene appears to have reduced the overall abundance of the animals available there while, at the same time, increasing the size and mobility of the herds in which these animals lived. It is also likely that the predictability of herd movements and seasonal aggregation patterns decreased throughout the early Holocene.

These changes should have increased the size of human aggregations but should also have decreased their regularity, especially after the end of the Folsom Period. A dispersed Clovis Period fauna should have led to a dispersed distribution of residential sites, whereas the likely importance of communal hunting and the larger and relatively predictable herds existing during Folsom times should have led hunters to repeatedly aggregate in larger groups in certain preferred locations. The decreasing predictability of the herds after Folsom times should have forced hunters to use a more diverse set of locations. The increasing unpredictability of the herds throughout the early Holocene should also have reduced the ability of groups of people to plan aggregations in advance and therefore should have reduced the regularity of those aggregations. In response to these changes, social complexity should have increased from Clovis to Folsom times and decreased thereafter. The available archival and published data on Paleoindian adaptations on the Southern High Plains and elsewhere are consistent with these predictions, although these data do not provide a rigorous test of them.

1. FUTURE RESEARCH ON PALEOINDIAN ORGANIZATION

The ultimate goal of this analysis was to derive a testable set of hypotheses about Paleoindian adaptations on the Southern High Plains that are consistent with the available data and that provide a basis for future research. The hypotheses considered here suggest several specific areas for such research.

Faunal remains and flaked stone tools and their manufacturing debris comprise the overwhelming majority of the remains recovered from Paleoindian sites on the Great Plains. The discussion of the structure of communal kill sites in Chapter 11 indicates one obvious avenue for research: excavations of such sites should examine the structure of the bone beds encountered to look for evidence of cooperative butchering. Although virtually all analyses of communal kills routinely report the evidence for herd composition, season of kill, and butchering techniques, detailed discussions of the distribution of bone within the sites are relatively rare.

Second, studies of stone tool production and use may also offer insights into some aspects of social complexity. The ability of a society to support full- or part-time craft specialists is one measure of that society's complexity, and the excellence of the stoneworking evident in the artifact assemblages recovered from Paleoindian sites suggests a degree of specialization in tool production. Hester and Grady (1977:94) and Knudson (1983:39, 99), for example, suggest that no more than two or three individuals, and possibly only one, produced the points found at each of several Paleoindian sites on the Plains. Considering Paleoindian projectile points in light of the possibility that they were produced by at least part-time specialist knappers is interesting in light of the changes in organizational complexity over time proposed in Chapter 11.

Although it may not be a universal characteristic of specialized production, it is likely that artifacts that are more difficult to produce are more likely to have been made by a more restricted number of individuals in a society than artifacts that are easier to produce. From this perspective, Paleoindian points can be divided into those finished by fluting (Clovis and Folsom points) and those finished by pressure flaking (later Paleoindian points). Experiments by modern flint knappers have demonstrated that fluting is by far the more difficult of these two techniques to master and that the process of producing a fluted point is far more complex than that of producing other types of points (i.e., Callahan 1979; Crabtree 1968; Flenniken 1978; Tunnell 1977; and many others). Modern apprentice knappers certainly become adept at pressure flaking well before they can successfully flute a preform.

Clovis and Folsom points, however, do not appear to be equally difficult to flute. The channel flake driven from a Clovis point generally extends no more than one-third to one-half of the distance down the point, and many Clovis points show evidence of multiple flutes on each face (Wormington 1957:263). Hester (1972:97) further describes fluted Clovis points from Blackwater Draw Locality No. 1 as "typically thick and heavy." Folsom points, in contrast, are thin, well-flaked, and were usually fluted by driving a single channel flake along most of their length (Wormington 1957:263). Although successfully fluting a Folsom point seems to require special tools (i.e., Akerman and Fagan 1986; Frison and

Bradley 1982; Tunnell 1977), one accomplished flint knapper (W. Jonas, personal communication) has suggested that some individuals can flute Clovis points while holding the preform in their hands.

These data thus imply that both Clovis and Folsom points are more difficult to produce than later points finished by pressure flaking and that Folsom points are the more difficult of the two fluted types to manufacture. If a reliance on tools that are more difficult to produce is indeed an indicator of a higher degree of specialized production in a society, this pattern corresponds to the expected increase in social complexity from Clovis to Folsom times and the likely decrease thereafter. The well-known "demise" of high-quality stoneworking at the end of the Paleoindian period (cf. Hayden 1982) may reflect the need for many or most of the members of a society to produce their own tools as the onset of an essentially modern climate and transition to the Altithermal created conditions under which a more complex social structure was difficult to maintain. The evidence for cooperative butchering of the animals taken in communal kills that was noted in Chapter 11 shows a similar pattern.

Testing for the existence of some form of specialized production during the Paleoindian Period is a major area for future research. Experimental verification of the relative difficulty of producing the different types of Paleoindian points is needed. There is some dispute at present about how often Folsom points break during production, with Frison and Bradley (1982:211) emphasizing experimental evidence of failure rates as high as 40% and Akerman and Fagan (1986) presenting a fluting technique with an 84% success rate that they claim less skilled knappers can use. Sollberger (1986), however, argues that the points Akerman and Fagan produced with this technique are not true replicas of prehistoric Folsom points and that more work is required to verify their success rate. Quantifying rates of artifact breakage during manufacture and the time needed for apprentices to master different modes of tool production would provide a more objective basis for inferring how difficult certain types of artifacts are to produce, providing a test of the arguments outlined above.

Second, the study of assemblages of Paleoindian tools and production debris, particularly assemblages recovered from campsites, should provide some insights into this issue. Individual structures have been identified at several Folsom period camps, including the Lindenmeier (Wilmsen and Roberts 1978), Agate Basin (Frison and Stanford 1982), and Hanson (Frison and Bradley 1981) sites. Studies of differences in the production debris associated with different structures in such sites could show the degree to which production of flaked stone tools was specialized within a residential unit: if certain classes of artifacts, such as projectile points, were produced primarily by the residents of certain structures, for example, some degree of specialization in stoneworking within a social group might be inferred.

Other aspects of stone tool production and use are also relevant to tests of the hypotheses outlined here. Procurement of raw material may have become less predictable as patterns of human movement became more irregular, and a greater reliance on local lithic resources could have resulted. The extensive Paleoindian use of low-quality local stone at Lubbock Lake (Bamforth 1985), for example, may reflect the relative paucity of Folsom-age lithic material from the site, although the Folsom levels at the site did produce a number of tools made of such stone.

The pattern of extreme raw material conservation often noted at Paleoindian sites (Goodyear 1979) could pose difficulties for lithic analyses directed at some of the questions posed here because stone tools may only have been maintained rather than actually produced at many locations. Sites that are close to raw material sources, such as Hanson (Frison and Bradley 1981), however, seem to show less conservative use of the available stone and more primary tool production, and thus are better candidates for intensive analysis of this type. The Lake Theo Site, a Folsom and Late Paleoindian period camp near the Tecovas jasper quarries in Briscoe County, Texas (Harrison and Killen 1978), is a site on the Southern High Plains which could provide information which is particularly relevant to the hypotheses presented here.

2. CONCLUSIONS

Despite the good agreement this study found between theoretical expectations and the available data, it is important to remember that environmental factors such as those considered here do not simply determine human responses. Rather, they exert their effects on the range of behavior available to a society trying to make a living in a region, and this range derives largely from that society's historical background. This relationship is perhaps clearest among the recent societies considered in Chapters 7 and 8. Oliver (1962) has shown that organizational differences among these groups can be traced back to the nature of the societies in the regions from which they migrated to the Plains, and the present study shows how these differences affected the ability of the migrating tribes to cope with environmental conditions in different parts of the region. Understanding the effects of historical forces resulting from the European expansion across North America is critical to understanding the Plains societies confronting that expansion, but these forces operated in and interacted with an environmental context with its own effects on society.

This analysis rests on a substantial body of research on grassland and ungulate ecology. Without the knowledge that this research provides, the relationships discussed here could not have been discerned. One point that this study should therefore illustrate is that in order to understand how human beings cope with environmental conditions, anthropologists need to devote as much attention to these conditions as to human responses. The question of the "predictability" of bison herding and migration patterns discussed in Chapter 6, for example, cannot be resolved without understanding the environmental factors that determine these patterns. Taking such factors into account, bison appear to be as predictable as any other large ungulate (also see Bamforth 1987). In many cases, the key aspects of the structure of a region's resource base will simply not be apparent without a detailed study of the climatic and other conditions that govern that structure.

The anthropological analyses here have produced two important conclusions. The first is that a clear relationship can be shown between organizational variability among the recent Plains tribes and the nature of the environments they inhabited. Second, there appears to be evidence for substantial changes in human adaptations over the course of the Paleoindian Period, at least on the Southern High Plains, and perhaps elsewhere. This conclusion differs from the traditional view that these adaptations varied primarily in the kinds of animals hunted and the types of projectile points used to hunt them (especially see Wendorf and Hester 1962), although other recent work, such as that discussed in Chapter 10, has reached similar conclusions. However, the inferences presented here about Paleoindian adaptations remain tentative because of the limits on the available data. This study's predictions about Paleoindian land-use patterns and organization appear to be consistent with these data, but they require substantially more extensive testing.

These conclusions rest on a theory that proposes that as environmental conditions, specifically a patchier and more predictable resource base, tend to favor larger and more regular human aggregations, human society should become more complex. Unlike many other discussions of social change, this theory does not necessarily rely on changes in overall population density or any other form of population/resource imbalance. Rather, it depends on how changes in the spatial and temporal distribution of resources within a region affect the number of people in face-to-face contact with one another and the length of time over which they maintain that contact. This theory predicts changes in land-use patterns and organizational complexity in a region even if resource abundance and human population size remain stable, if changes in the structure of the resources in an area favored larger, more regular, or longer human aggregations. Larger populations within a region may often increase local group size and thus trigger the processes discussed here, but such increases are not inevitable.

The generality of the theory presented here makes it potentially applicable in many times and places other than those to which this study applies it. Translating the general predictions of this theory for specific times or places other than the Great Plains, however. will require reliable information on the nature and distributions of the different resources in them. Attempts to compile such information should recognize the clear importance of the fact that no human group lives in an environment that is exactly the same from year to year: The analysis in Chapter 8 in particular should show that average environmental conditions are not necessarily the major determinants of human adaptations, and anthropological research into the relationship between human beings and the environments they inhabit should take this fact into account. The discussion here should show one example of how this can be done, as well as the strength of the cultural–ecological relationships it can reveal.

References

Aandahl, A.R., 1982, *Soils of the Great Plains,* University of Nebraska Press, Lincoln.

Agenbroad, L.D., 1978, *The Hudson-Meng Site: An Alberta Bison Kill in the Nebraska High Plains,* University Press of America, Washington, DC.

Agogino, G., 1969, The Midland Complex—Is It Valid? *American Anthropologist* 71:1117–1118.

Akerman, K., and Fagan, J.L., 1986, Fluting the Lindenmeier Folsom: A Simple and Economical Solution to the Problem, and Its Implications for Other Fluted Point Technologies, *Lithic Technology* 15:1–7.

Akin, D.E., and Burdick, D., 1977, Rumen Microbial Degradation of Starch-Containing Bundle Sheath Cells in Warm-Season Grasses, *Crop Science* 17:529–533.

Allred, B.W., and Mitchell, H.C., 1955, Major Plant Types of Arkansas, Louisiana, Oklahoma, and Texas and Their Relation to Climate and Soils, *Texas Journal of Science* 7:7–19.

Andrewartha, H.G., 1961, *An Introduction to the Study of Animal Populations,* University of Chicago Press, Chicago.

Arnold, G.W., and Dudzinski, M.L., 1978, *Ethology of Free-Ranging Domestic Animals,* Elsevier Scientific Publishing Company, New York.

Arthur, G.A., 1975, An Introduction to the Ecology of Early Historic Communal Bison Hunting among the Northern Plains Tribes, *Archaeological Survey of Canada Paper* 37, National Museums of Canada, Ottawa.

Audubon, J., and Bachman, J., 1854, *The Quadrupeds of North America,* V.G. Audubon, New York.

Ayeni, J.S.O., 1975, Utilization of Waterholes in Tsavo National Park (East), *East African Wildlife Journal* 13:305–323.

Baker, R.R., 1978, *The Evolutionary Ecology of Animal Migration,* Hodder and Staughton, London.

Bamforth, D., 1985, The Technological Organization of Paleoindian Small-Group Bison Hunting on the Llano Estacado, *Plains Anthropologist* 30:243–258.

Bamforth, D., 1987, Historical Documents and Bison Ecology on the Great Plains, *Plains Anthropologist,* 32:1–16.

Barrett, M.W., 1982, Distribution, Behavior, and Mortality of Pronghorns during a Severe Winter in Alberta, *Journal of Wildlife Management* 46:991–1002.

Bedord, J.N., 1974, Morphological Variation in Bison Metacarpals and Metatarsals, in: *The Casper Site* (George Frison, ed.), Academic Press, New York, pp. 199–240.

Bell, R.H.V., 1971, A Grazing Ecosystem in the Serengeti, *Scientific American* 225:86–93.

Bettinger, R.L., 1978, Alternative Adaptive Strategies in the Prehistoric Great Basin, *Journal of Anthropological Research* 34:27–46.

Binford, L.R., 1980, Willow Smoke and Dog's Tails: Hunter-Gatherer Settlement Systems and Archaeological Site Formation, *American Antiquity* 45:4–20.

Binford, L.R., 1982, The Archaeology of Place, *Journal of Anthropological Archaeology* 2:15–33.

Binford, L.R., 1983, *Nunamiut Ethnoarchaeology,* Academic Press, New York.

Blair, W.F., 1950, The Biotic Provinces of Texas, *Texas Journal of Science* 2:93–117.

Blasing, T.J., and Fritts, H.C., 1976, Reconstructing Past Climatic Anomalies in the North Pacific and Western North America from Tree-Ring Data, *Quaternary Research* 6:563–579.

Bokhari, U.G., 1978a, Total Nonstructural Carbohydrates in the Vegetation Components of a Shortgrass Prairie Ecosystem under Stress Conditions, *Journal of Range Management* 31:224–230.

Bokhari, U.G., 1978b, Nutritional Characteristics of Blue Grama Herbage under the Influence of Added Water and Nitrogen, *Journal of Range Management* 31:18–22.

Bolton, H.E., 1908, *Spanish Exploration in the Southwest, 1542–1706,* Barnes and Noble, New York.

Borchert, J.R., 1950, The Climate of the North American Grasslands, *Annals of the Association of American Geographers* 40:1–39.

Boutton, T.W., Harrison, A.T., and Smith, B.N., 1980, Distribution of Biomass of Species Differing in Photosynthetic Pathway along an Altitudinal Transect in Southeastern Wyoming Grassland, *Oecologia* 45:287–298.

Bradley, R.S., 1985, *Quaternary Paleoclimatology,* Allen and Unwin, Boston.

Bray, E.C., and Bray, M.C., 1976, *Joseph N. Nicollet on the Plains and Prairies,* Historic Society Press, St. Paul, Minnesota.

Brink, J., 1986, Dog Days on the Alberta Prairies, *Occasional Papers of the Alberta Archaeological Society* 9.

Brody, S., 1945, *Bioenergetics and Growth,* Hafner Publishing Co., New York.

Brown, R.H., 1978, A Difference in N Use Efficiency in C3 and C4 Plants and its Implications in Adaptation and Evolution, *Crop Science* 18:93–98.

Brown, R.W., 1977, Water Relations of Range Plants, in: *Rangeland Plant Physiology* (R. Sosebee, ed.), Society for Range Management, Denver, pp. 97–140.

Bryant, E.H., 1973, Habitat Selection in a Variable Environment, *Journal of Theoretical Biology* 41:421–429.

Bryant, V.M., and Shafer, H.J., 1977, The Late Quaternary Paleoenvironment of Texas: A Model for the Archaeologist, *Texas Archaeological Society Bulletin* 48:1–25.

Bryson, R.A., and Wendland, W.A., 1969, Tentative Climatic Patterns for Some Late Glacial and Post-Glacial Episodes in Central North America, in: *Life, Land, and Water* (W.J. Mayer-Oakes, ed.), University of Manitoba Press, Winnipeg, pp. 271–298.

Bryson, R.A., Baerreis, D.A., and Wendland, W.M., 1970, The Character of Late-Glacial and Post-Glacial Climatic changes, in: *Pleistocene and Recent Environments of the Central Great Plains* (W. Dort and J.K. Jones, eds.), University of Kansas Press, Lawrence, Kansas, pp. 53–74.

Buwai, M., and J. Trlica, 1977, Multiple Defoliation Effects on Herbage Yield, Vigor, and Total Nonstructural Carbohydrates of Five Range Species, *Journal of Range Management* 30:164–171.

Byers, D.S., 1954, Bull Brook: A Fluted Point Site in Ipswich, Massachusetts, *American Antiquity* 19:343–351.

Callahan, E., 1979, The Basics of Biface Knapping in the Eastern Fluted Point Tradition, *Archaeology of Eastern North America* 7:1–180.

Carneiro, R.L., 1967, On the Relationship between Size of Population and Complexity of Social Organization, *Southwestern Journal of Anthropology* 23:234–243.

Carneiro, R.L., 1970, A Theory of the Origin of the State, *Science* 169:733–738.

Caswell, H., Reed, F., Stephenson, S.N., and Werner, P.A., 1973, Photosynthetic Pathways and Selective Herbivory: A Hypothesis, *American Naturalist* 107:465–480.

Church, D.C., 1971, *Digestive Physiology and Nutrition of Ruminants,* D.C. Church and the Oregon State University Book Stores, Inc., Corvallis, Oregon.

Coe, M.J., Cummings, D.H., and Phillipson, J., 1976, Biomass and Production of Large African Herbivores in Relation to Rainfall and Primary Production, *Oecologia* 22:341–354.

Cogswell, C., and Kamstra, L.D., 1976, The Stage of Maturity and Its Effect upon the Chemical Composition of Four Native Range Species, *Journal of Range Management* 29:460–463.

Cole, C.V., Innis, G.S., and Stewart, J.W.B., 1978, Simulation of Phosphorus Cycling in Semiarid Grasslands, in: *Grassland Simulation Model* (G.S. Innis, ed.), Springer-Verlag, New York, pp. 205–230.

Colinvaux, P., 1973, *Introduction to Ecology,* Wiley, New York.

Colinvaux, P., 1980, *The Fates of Nations,* Simon and Schuster, New York.

Colson, E., 1954, Review of "The Comanche: Lords of the South Plains," by H. Wallace and E. Hoebel, *Man* 54:13–14.

Colwell, R.K., 1974, Predictability, Constancy, and Contingency of Periodic Phenomena, *Ecology* 55:1148–1153.

Conkey, M., 1978, Style and Information in Cultural Evolution: Toward a Predictive Model for the Pleistocene, in: *Social Archaeology: Beyond Subsistence and Dating* (C.L. Redman ed.), Academic Press, New York, pp. 61–85.

Conkey, M., 1980, The Identification of Prehistoric Aggregation Sites: The Case of Altamira, *Current Anthropology* 21:609–630.

Cooper, J.M., 1956, The Gros Ventres of Montana, Part 2: Religion and Ritual, *Catholic University of America Anthropological Series* 16.

Coppock, D.L., Detling, J.K., Ellis, J.E., and Dyer, M.I., 1983a, Plant-Herbivore Interactions in a North American Mixed-Grass Prairie I, *Oecologia* 56:1–9.

Coppock, C.L., Detling, J.D., Ellis, J.E., and Dyer, M.I., 1983b, Plant-Herbivore Interactions in a North American Mixed-Grass Prairie II, *Oecologia* 56:10–15.

Cordell, L., and Plog, F., 1979, Escaping the Confines of Normative Thought: A Reevaluation of Puebloan Prehistory, *American Antiquity* 44:405–429.

Coues, E., 1897, *New Light on the Early History of the Great Northwest,* F.P. Harper, New York.

Coupland, R.T., 1958, The Effects of Fluctuations in Weather upon the Grasslands of the Great Plains, *The Botanical Review* 24:273–317.

Coupland, R.T., 1959, Effects of Changes in Weather Conditions upon Grasslands in the Northern Great Plains, in: *Grasslands* (H.B. Sprague, ed.), American Association for the Advancement of Science, Washington, pp. 291–306.

Coupland, R.T., 1961, A Reconsideration of Grassland Classification in the Northern Great Plains of North America, *Journal of Ecology* 49:135–167.

Coupland, R.T., 1979, Distribution of Grasses and Grasslands of North America, in: *Ecology of Grasslands and Bamboolands in the World* (M. Numata, ed.), Mouton, Boston, pp. 77–83.

Court, A., 1974, The Climate of the Coterminous United States, in: *Climates of North America* (R.A. Bryson and F.K. Hare, eds.), Elsevier Scientific Publishing Co., New York, pp. 193–344.

Cowgill, G.L., 1977, The Trouble with Significance Tests and What We Can Do About It, *American Antiquity* 42:350–368.

Crabtree, D.E., 1966, A Stone-Worker's Approach to Analyzing and Replicating the Lindenmeier Folsom, *Tebiwa* 9:3039.

Crowder, L.V., and Chheda, H.R., 1982, *Tropical Grassland Husbandry,* Longman, New York.

Culbertson, T.A., 1952, Journal of an Expedition to the Mauvaises Terres and the Upper Missouri in 1850, *Bureau of American Ethnology Bulletin* 147.

Davis, E.M., 1953, Recent Data from Two Paleoindian Sites on Medicine Creek, Nebraska, *American Antiquity* 18:380–386.

Davis. L., and Wilson, M., 1978, Bison Procurement and Utilization: A Symposium, *Plains Anthropologist Memoir* 14.

Delaney, M.J., and Happold, D.C.C., 1979, *Ecology of Large African Mammals,* Longman Group Limited, New York.

Denig, E.T., 1930, Indian Tribes of the Upper Missouri, *Bureau of American Ethnology Annual Report* 46.

Denig, E.T., 1961, *Five Indian Tribes of the Upper Missouri,* University of Oklahoma Press, Norman.

Dibble, D., 1965, Bonfire Shelter: A Stratified Bison Kill Site in the Amistad Reservoir Area, Val Verde County, Texas, *Texas Archaeological Salvage Project Miscellaneous Papers* 5.

Dickinson, C.E., and Dodd, J.L., 1976, Phenological Pattern in the Shortgrass Prairie, *American Midland Naturalist* 96:367–378.

Dodge, R.I., 1884, *Our Wild Indians,* A.D. Worthington, Hartford, Connecticut.

Dodge, R.I., 1965, *The Black Hills,* Ross and Haines, Minneapolis.

Dorsey, J.O., 1891, The Social Organization of the Siouan Tribes, *Journal of American Folklore* 4:257–266.

Dorsey, J.O., 1905, The Cheyenne. *Field Columbian Museum Anthropological Series* 9, Nos. 1 and 2.

Dudnik, E., 1971, *SYMAP,* Department of Architecture, University of Illinois, Chicago.

Duvick, D., and Blasing, T., 1981, A Dendroclimatic Reconstruction of Annual Precipitation Amounts in Iowa since 1680, *Water Resources Research* 17:1183–1189.

Dwyer, P.D., and Minnegal, M., 1985, Andaman Islanders, Pygmies, and an Extension of Horn's Model, *Human Ecology* 13:111–119.

Eck, H.V., McGully, W.G., and Stubbendieck, J., 1975, Response of Shortgrass Plains Vegetation to Clipping, Precipitation, and Soil Water, *Journal of Range Management* 28:194–197.

Eggan, F., 1952, The Ethnological Cultures and Their Archaeological Backgrounds, in: *The Archaeology of the Eastern United States* (J.B. Griffin, ed.), University of Chicago Press, Chicago, pp. 35–45.

Eltringham, S.K., 1974, Changes in the Large Mammal Community of Mweya Peninsula, Rwenzori National Park, Uganda, Following Removal of Hippopatamus, *Journal of Applied Ecology* 11:855–866.

Ember, M., 1963, The Relationship between Economic and Political Development in Nonindustrialized Societies, *Ethnology* 2:228–248.

Evans, G.L., 1951, Prehistoric Wells in Eastern New Mexico, *American Antiquity* 17:1–9.

Everitt, H., Alaniz, M.A., and Gerberman, A.H., 1982, Chemical Composition of Native Range Grasses Growing on Saline Soils of the South Texas Plains, *Journal of Range Management* 35:43–46.

Ewers, J.C., 1949, The Last Bison Drives of the Blackfoot Indians, *Journal of the Washington Academy of Science.* 39.

Ewers, J.C., 1955, *The Horse in Blackfoot Indian Culture,* Smithsonian Institution Press, Washington, DC.

Fawcett, W.B., 1986, Variation in the Lithic Assmblages from Bison Kills and Associated Camp and Processing Areas, *Wyoming Archaeologist* 29:9–28.

Fawcett, W.B., 1987, *Communal Hunts, Human Aggregations, Social Variation, and Climatic Change: Bison Utilization by Prehistoric Inhabitants of the Great Plains,* PhD Dissertation, Department of Anthropology, University of Massachusetts, Amherst.

Fenneman, N.M., 1938, *Physiography of the Eastern United States,* McGraw-Hill, New York.

Fiedler, A.G., and Nye, S.S., 1933. Geology and Ground-Water Resources of the Roswell Artesian Basin, New Mexico, *U.S. Geological Survey Water Supply Paper* 639, United States Government Printing Office, Washington, DC.

Figgins, J.D., 1927, The Antiquity of Man in America, *Natural History* 3:229–239.

Finley, R.J., and Gustavson, T., 1980, Climatic Controls on Erosion in the Rolling Plains and along the Caprock Escarpment of the Texas Panhandle, *University of Texas at Austin Bureau of Economic Geology Geological Circular* 80–11.

Flannery, R., 1953, The Gros Ventres of Montana, Part 1: Social Life, *Catholic University of American Anthropological Series* 15.

Flenniken, J.F., 1978, Reevaluation of the Lindenmeier Folsom: A Replication Experiment in Lithic Technology, *American Antiquity* 43:473–480.

Forbis, R.G., and Sperry, J.D., 1952, An Early Man Site in Montana, *American Antiquity* 18:127–133.

Ford, R.I., 1972, Barter, Gift, or Violence: An Analysis of Tewa Intertribal Exchange, in: Social Exchange and Interaction (E. Wilmsen. ed.), *University of Michigan Museum of anthropology Anthropological Paper* 46, pp. 21–45.

Fowler, L., 1982, "Look at My Hair, It is Grey": Age Grading, Ritual Authority, and Political Change among the Northern Arapahoes and Gros Ventres, *Smithsonian Contributions to Anthropology* 30:73–93.

French, N.R., 1979, Introduction, in: *Grassland Ecosystems of the World* (R.T. Coupland, ed.), Cambridge University Press, New York, pp. 41–48.

Frison, G.C., 1967, *Archaeological Evidence of the Crow Indians in Northern Wyoming: A Study of a Late Prehistoric Buffalo Economy,* PhD Dissertation, Anthropology Department, University of Michigan.

Frison, G.C., 1971, The Buffalo Pound in Northwestern Plains Prehistory: Site 48CA302, Wyoming, *American Antiquity* 36:77–91.

Frison, G.C., 1972, The Role of Buffalo Procurement in Post-Altithermal Populations on the Northwestern Plains, *University of Michigan Museum of Anthropology Anthropological Papers* 46:11–20.

Frison, G.C., 1973, The Wardell Buffalo Trap 48SU301: Communal Procurement in the Upper Green River Basin, Wyoming, *University of Michigan Museum of Anthropology Anthropological Papers* 48.

Frison, G.C., 1974, *The Casper Site,* Academic Press, New York.

Frison, G.C., 1976, Crow Pottery in Northern Wyoming, *Plains Anthropologist* 21:29–44.

Frison, G.C., 1978, *Prehistoric Hunters of the High Plains,* Academic Press, New York.

Frison, G.C., 1982, Paleoindian Winter Subsistence Strategies on the High Plains, *Smithsonian Contributions to Anthropology* 30:194–201.

Frison, G.C., 1984, The Carter/Kerr-McGee Paleoindian Site: Cultural Resource Management and Archaeological Research, *American Antiquity* 49:288–314.

Frison, G.C., and Bradley, B.A., 1981, *Folsom Tools and Technology at the Hanson Site, Wyoming,* University of New Mexico Press, Albuquerque.

Frison, G.C., and Bradley, B.A., 1982, Fluting Folsom Projectile Points, in: *The Agate Basin Site* (G. Frison and D. Stanford, eds.), Academic Press, New York, pp. 209–212.

Frison, G.C., and Stanford, D., 1982, *The Agate Basin Site: A Record of Paleoindian Occupation of the Northwestern Plains,* Academic Press, New York.

Frison, G.C., Walker, D., Webb, S., and Ziemens, G., 1977, Paleoindian Procurement of Camelops on the Northwestern Plains, *Quaternary Research* 10:385–400.

Fritts, H.C., Lofgren, G.R., and Gordon, G.A., 1979, Variations in Climate since 1602 as Reconstructed from Tree Rings, *Quaternary Research* 12:18–46.

Fritts, H.C., Lofgren, G.R., and Gordon, G.A., 1981, Reconstructing Seasonal to Century Time Scale Variations in Climate from Tree-Ring Evidence, in: *Climate and History* (T.M.L. Wigley, M.J. Ingram, and G. Farmer, eds.), Cambridge University Press, New York, pp. 139–161.

Fulgham, T., and Stanford, D., 1982, The Frasca Site: A Preliminary Report, *Southwestern Lore* 48:1–9.

Fuller, W.A., 1960, Behavior and Social Organization of the Wild Bison of Wood Buffalo National Park, Canada, *Arctic* 13:3–19.

Fuller, W.A., 1961, The Ecology and Management of the American Bison, *Terre et Vie* 5:286–304.

Goetz, H., 1975, Effect of Site and Fertilization on Protein Content of Native Grasses, *Journal of Range Management* 28:380–385.

Goodyear, A.C., 1979, A Hypothesis for the Use of Cryptocrystalline Raw Materials among Paleo-Indian Groups of North America, *University of South Carolina Research Manuscript Series* 156.

Gorman, F., 1972, The Clovis Hunters: An Alternate View of Their Environment and Ecology, in: *Contemporary Archaeology* (M. Leone, ed.), Southern Illinois University Press, Carbondale, pp. 206–222.

Gould, C.N., 1906, The Geology and Water Resources of the Eastern Portion of the Panhandle of Texas, *U.S. Geological Survey Water Supply Paper* 154, United States Government Printing Office, Washington, DC.

Graham, R.W., and Lundelius, E.L., 1984, Coevolutionary Disequilibrium and Pleistocene Extinctions, in: *Quaternary Extinctions: A Prehistoric Revolution* (P. Martin and R. Klein, eds.), University of Arizona Press, Tucson, pp. 223–249.

Gramly, R., 1982, The Vail Site: A Paleoindian Encampment in Maine, *Buffalo Museum of Sciences Bulletin* 30.

Green, F.E., 1961, The Monahans Dunes Area, in: *Paleoecology of the Llano Estacado* (F. Wendorf, Ed.), Fort Burgwin Research Center Publication No. 1, pp. 22–47.

Greiser, S.T., 1985, Predictive Models of Hunter–Gatherer Subsistence and Settlement Strategies on the Central High Plains, *Plains Anthropologist Memoir* 20.

Grimes, J.R., Eldridge, W., Grimes, B.G., Vaccaro, A., Vacarro, F., Vaccaro, J., Vaccaro, N., and Orsini, A., 1984, Bull Brook II, *Archaeology of Eastern North America* 12:159–183.

Grinnell, G.B., 1962a, *The Cheyenne Indians, Vol. 1,* Cooper Square Publishers, New York.

Grinnell, G.B., 1962b, *The Cheyenne Indians, Vol. 2,* Cooper Square Publishers, New York.

Grobler, J.H., and Jones, M.A., 1980, Population Statistics and Carrying Capacity of Large Ungulates in the Whovi Wild Area, Rhodes Matapos National Park, Zimbabwe, Rhodesia, *South African Journal of Wildlife Research* 10:38–42.

Guffee, E.J., 1979, *The Plainview Site: Relocation and Archaeological Investigation of a Late Paleo-Indian Kill Site in Hale County, Texas,* Llano Estacado Museum, Plainview.

Gunnerson, D., 1974, *The Jicarilla Apaches,* Northern Illinois University Press, Dekalb.

Guthrie, R.D., 1980, Bison and Man in North America, *Canadian Journal of Anthropology* 1:55–73.

Guthrie, R.D., 1984, Mosaics, Allelochemics, and Nutrients: An Ecological theory of Late Pleistocene Extinctions, in: *Quaternary Extinctions: A Prehistoric Revolution* (P. Martin and R. Klein, eds.), University of Arizona Press, Tucson, pp. 259–298.

Haines, F., 1938a, Where did the Plains Indians Get Their Horses? *American Anthropologist* 40:112–117.

Haines, F., 1938b, The Northward Spread of Horses among the Plains Indians, *American Anthropologist* 40:429–437.

Hall, R., 1973, *Wild Horse: Biology for Management,* Bureau of Land Management, Billings.

Hamilton, W.D., 1971, Geometry for the Selfish Herd, *Journal of Theoretical Biology* 31:295–311.

Hammond, G.P., and Rey, A., 1940, *Narratives of the Coronado Expedition, 1540–1542,* University of New Mexico Press, Albuquerque.

Hammond, G.P., and Rey, A., 1966, *The Rediscovery of New Mexico, 1580–1594,* University of New Mexico Press, Albuquerque.

Hanson, J.R., 1984, Bison Ecology in the Northern Plains and a Reconstruction of Bison Patterns for the North Dakota Region, *Plains Anthropologist* 29:93–113.

Haragan, D.R., 1976, Spatial Variation of Precipitation on the Texas High Plains, *Water Resources Bulletin* 12:1191–1204.

Hare, F.K., and Hay, J.E., 1974, The Climate of Canada and Alaska, in: *Climates of North America* (R.A. Bryson and F.K. Hare, eds.), Elsevier Scientific Publishing Co., New York, pp. 49–192.

Harrison, B., and Smith, H., 1975, A Test Excavation of the Lake Theo Site, Briscoe County, Texas, *Panhandle-Plains Historical Review* 48:70–108.

Harrison, B.R., and Killen, K., 1978, Lake Theo: A Stratified Early Man Bison Butchering and Camp Site, Briscoe County, Texas, *Panhandle-Plains Historical Museum Special Archaeological Report* 1.

Hart, R.H., Abdalla, O.M., Clark, D.H., Marshall, M.B., Hamid, M.H., Hagar, J.A., and Waggoner, Jr., J.W., 1983, Quality of Forage and Cattle Diets on the Wyoming High Plains, *Journal of Range Management* 36:46–51.

Hassrick, R.B., 1964. *The Sioux,* University of Oklahoma Press, Norman.

Hayden, B., 1982, Interaction Parameters and the Demise of Paleoindian Craftsmanship, *Plains Anthropologist* 27:109–123.

Hayden, F.V., 1862, Contributions to the Ethnography and Philology of the Indian Tribes of the Missouri Valley, *Transactions of the American Philosophical Society* 12:231–461.

Haynes, C.V., 1975, Pleistocene and Recent Stratigraphy, in: *Late Pleistocene Environments of the Southern High Plains* (F. Wendorf and J. Hester, eds.), Fort Burgwin Research Center Publication no. 9, pp. 57–96.

Haynes, C.V., 1980, The Clovis Culture, *Canadian Journal of Anthropology* 1:115–121.

Heffly, S., 1981, The Relationship between Northern Athapaskan Settlement Patterns and Resource Distribution: An Application of Horn's Model, in: *Hunter Gatherer Foraging Strategies* (B. Winterhalder and E. Smith, eds.), University of Chicago Press, Chicago, pp. 126–147.

Hemmings, E.T., 1970, *Early Man in the San Pedro Valley, Arizona,* PhD Dissertation, Anthropology Department, University of Arizona.

Hennessy, J.T., Gibbens, R.P., Tromble, J.M., and Cardenas, M., 1983, Vegetation Changes from 1935 to 1980 in Mesquite Dunelands and Former Grasslands of Southern New Mexico, *Journal of Range Management* 36:370–374.

Herschfeld, D.M., 1962, A Note on the Variability of Annual Precipitation, *Journal of Applied Meteorology* 1:575–578.

Hester, J., 1961, The Elida Site, Evidence of a Folsom Occupation in Roosevelt County, Eastern New Mexico, *Texas Archaeological Society Bulletin* 30:209–215.

Hester, J., 1962, A Folsom Lithic Complex from the Elida Site, Roosevelt County, N.M., *El Palacio* 12:92–113.

Hester, J., 1972, Blackwater Draw Locality No. 1: A Stratified Early Man Site in Eastern New Mexico, *Fort Burgwin Research Center Publication* 8.

Hester, J., 1975, Paleoarchaeology of the Llano Estacado, in: *Late Pleistocene Environments of the Southern High Plains* (F. Wendorf and J. Hester, eds.), Fort Burgwin Research Center Publication no. 9, pp. 247–256.

Hester, J., and Grady, J., 1977, Paleoindian Social Patterns on the Llano Estacado, in *Paleoindian Lifeways* (E. Johnson, ed.), West Texas Museum Association, Lubbock, pp. 78–96.

Hewes, G., 1948, Early Tribal Movements in the Northern Great Plains, *Plains Archaeological Conference Newsletter* 1:49–61.

Hills, L.V., 1969, Problems in Palynology: A Discussion, in: *Post-Pleistocene Man and His Environment on the Northern Plains* (M. Wilson, ed.), University of Calgary Archaeological Association, pp. 67–107.

Hobbs., M., Barker, D., and Gill, R., 1983, Comparative Nutritional Ecology of Montane Ungulates during Winter, *Journal of Wildlife Management* 47:1–16.

Hodder. I., 1979, Economic and Social Stress and Material Culture Patterning, *American Antiquity* 44:446–454.

Hodge, F.W. (ed.), 1907, Handbook of American Indians North of Mexico, Part 1, *Bureau of American Ethnology Bulletin* 30.

Hodge, F.W. (ed.), 1910, Handbook of North American Indians North of Mexico, Part 2, *Bureau of American Ethnology Bulletin* 33.

Hoebel, E.A., 1978, *The Cheyenne,* Little, Brown, New York.

Hoebel, E.A., 1980, On Cheyenne Socio-political Organization, *Plains Anthropologist* 25:161–170.

Hoijer, H., 1956, The Chronology of the Athapaskan Languages, *International Journal of American linguistics* 22:219–232.

Holliday, V.T., 1983, *Friends of the Pleistocene South-Central Cell 1983 Field Trip Guidebook to the Central Llano Estacado,* Texas Tech University Museum, Lubbock.

Holliday, V.T., 1985, New Data on the Stratigraphy and Pedology of the Clovis and Plainview Sites, Southern High Plains, *Quaternary Research* 23:388–402.

Holliday, V.T., 1986, Late Pleistocene Vegetation of the Southern High Plains: A Reappraisal. *Current Research in the Pleistocene* 3:53–54.

Holliday, V.T., and Johnson, E., 1981, An Update on the Plainview Occupation at the Lubbock Lake Site, *Plains Anthropologist* 26:251–253.

Holliday, V.T., Johnson, E., Hass, H., and Stuckenrath, R., 1983, Radiocarbon Ages from the Lubbock Lake Site 1950–1080: Framework for Cultural and Ecological Changes on the Southern High Plains, *Plains Anthropologist* 28:165–182.

Holliday, V.T., Johnson, E., Hall, S., and Bryant, V., 1985, Re-Evaluation of the Lubbock Subpluvial, *Current Research in the Pleistocene* 2:119–121.

Honea, K., 1980, Marks Beach, Stratified Paleoindian Site, Lamb County, Texas: Preliminary Report, *Texas Archaeological Society Bulletin* 51:243–269.

Horn, H.S., 1968, The Adaptive Significance of Colonial Nesting in the Brewer's Blackbird (Euphagus Cyanocephalus), *Ecology* 49:682–694.

Hornaday, W.T., 1889, The Extermination of the American Bison, with a Sketch of Its Discovery and Life History, *Smithsonian Report, 1887*:367–548.

Houston, D.B., 1982, *The Northern Yellowstone Elk Herd: Ecology and Management,* Macmillan Publishing Co., New York.

Hsiao, T.C., 1973, Plant Responses to Water Stress, *Annual Review of Plant Physiology* 24:519–570.

Hughes, J., and Willey, P., 1978, *Archaeology at Mackenzie Reservoir,* Texas Historical Commission, Austin.

Hughes, S.S., 1978, Bison Diminution on the Great Plains, *Wyoming Contributions to Anthropology* 1:18–47.

Hultkrantz, A., 1968, Shoshoni Indians on the Plains: An Appraisal of the Evidence, *Zeitschrift für Ethnologie* 93:4–72.

Hyde, G.E., 1937, *Red Cloud's Folk: A History of the Oglala Sioux Indians,* University of Oklahoma Press, Norman.

Hyde, G.E., 1959, *Indians of the High Plains,* University of Oklahoma Press, Norman.

Hyde, G.E., 1968, *Life of George Bent,* University of Oklahoma Press, Norman.

Ingram, M.J., Farmer, G., and Wigley, T.M.L., 1981, Past Climates and Their Impact on Man: A Review, in: *Climate and History* (T.M.L. Wigley, M.J. Ingram, and G. Farmer, Eds.), Cambridge University Press, New York, pp. 1–50.

Innis, G.S., 1977, *Grassland Simulation Model,* Springer-Verlag, New York.

Iowa Cooperative Wildlife Research Unit (ICWRU), 1980, *Range Ecology of Bison on Mixed Grass Prairie at Wind Cave National Park,* University of Wyoming-National Park Service Research Center, Laramie.

Irwin-Williams, C., Irwin, H., Agogino, G., and Haynes, C.V., 1973, Hell Gap: Paleoindian Occupation on the High Plains, *Plains Anthropologist* 19:40–53.

Jablow, J., 1951, The Cheyenne in Plains Indian Trade Relations 1795–1840, *American Ethnological Society Monograph* 19.

Jarman, P.J., 1974, The Social Organization of the Antelope in Relation to Their Ecology, *Behavior* 48:215–266.

Jarman, P.J., and Jarman, M.V., 1979, The Dynamics of Ungulate Social Organization, in: *Serengeti: Dynamics of an Ecosystem* (A.R.E. Sinclair and M. Norton-Griffiths, eds.), University of Chicago Press, Chicago, pp. 185–220.

Jarman, P.J., and Sinclair, A.R.E., 1979, Feeding Strategy and the Pattern of Resource Partitioning in Ungulates, in: *Serengeti: Dynamics of an Ecosystem* (A.R.E. Sinclair and M. Norton-Griffiths, eds.), University of Chicago Press, Chicago, pp. 130–163.

Jelinek, A.J., 1967, A Prehistoric Sequence in the Middle Pecos Valley, New Mexico, *University of Michigan Museum of Anthropology Anthropological Papers* 31.

Jenness, D., 1938, The Sarcee Indians of Alberta, *Canada Department of Mines and Resources Bulletin* 90.

Jenness, D., 1977, *The Indians of Canada,* National Museums of Canada, Ottawa.

Jewell, P.A., 1966, The Concept of Home Range in Mammals, in: Play, Exploration, and Territory in Mammals (P.A. Jewell and C. Loizos, eds.), *Symposia of the Zoological Society of London* 18, pp. 85–110.

Jochim, M.A., 1976, *Hunter-Gatherer Subsistence and Settlement: A Predictive Model,* Academic Press, New York.

Jochim, M.A., 1983, Optimization Models in Context, in: *Archaeological Theories and Hammers* (J.A. Moore and A. Keen, eds.), Academic Press, New York, pp. 157–172.

Johnson, E., 1977, Animal Food Resources of Paleoindians, in: *Paleoindian Lifeways* (E. Johnson, ed.), West Texas Museum Association, Lubbock, pp. 65–77.

Johnson, E., 1988a, Paleoenvironmental Overview, in: *Lubbock Lake: Late Quaternary Studies on the Southern High Plains,* Texas A and M University Press, in press.

Johnson, E., 1988b, *Lubbock Lake: Late Quaternary Studies on the Southern High Plains.* Texas A and M University Press, in press.

Johnson, E., and Holliday, V.T., 1980, A Plainview Kill-Butchering Locale on the Llano Estacado—The Lubbock Lake Site, *Plains Anthropologist* 25:89–111.

Johnson, E., and Holliday, V.T., 1981, Late Paleoindian Activity at the Lubbock Lake Site, *Plains Anthropologist* 26:173–193.

Johnson, E., Holliday, V.T., and Neck, R.W., 1982, Lake Theo: Late Quaternary Paleoenvironmental Data and New Plainview (Paleoindian) Date, *North American Archaeologist* 3:113–137.

Johnson, G.A., 1982, Organizational Structure and Scalar Stress, in: *Theory and Explanation in Archaeology* (C. Renfrew, M. Lowlands, and B. Segraves, eds.), Academic Press, New York, pp. 389–421.

Judge, W.J., n.d., Early Man: Plains and Southwest, in: *Handbook of North American Indians* (W.C. Sturvenant, ed.), *Vol.* 3, in press, Smithsonian Institution, Washington, DC.

Karl, K.R., and Koscielny, A.J., 1982, Drought in the United States: 1895–1981, *Journal of Climatology* 2:313–329.

Kehoe, T., 1973, The Gull Lake Site: A Prehistoric Bison Drive Site in Southwestern Saskatchewan, *Milwaukee Public Museum Publications in Anthropology and History* 1.

Kehoe, T., 1978, Paleoindian Bison Drives: Feasibility Studies, in: Bison Procurement and Utilization: A Symposium (L.B. Davis and M. Wilson, eds.), *Plains Anthropologist Memoir* 14, pp. 79–83.

Kelly, R.L., 1983, Hunter-Gatherer Mobility Strategies, *Journal of Anthropological Research* 39:277–306.

Kelly, R.L., and Todd, L.C., 1988, Coming into the Country: Early Paleoindian Hunting and Mobility, *American Antiquity* 53:231–244.

Kemp, P.R., and Williams, III, G.J., 1980, A Physiological Basis for Niche Separation between Agropyron Smithii (C3) and Bouteloua Gracilis (C4), *Ecology* 61:846–858.

Kennard, D., 1975, *Canadian Breaks,* University of Texas Division of Natural Resources and Environment, Austin.

Keyser, J.D., 1975, A Shoshonean Origin for the Plains Shield-Bearing Warrior Motif, *Plains Anthropologist* 20:207–215.

Keyser, J.D., 1977, Writing-on-Stone: Rock Art on the Northwestern Plains, *Canadian Journal of Archaeology* 1:15–80.

Keyser, J.D., 1979, The Plains Indian War Complex and the Rock Art of Writing-On-Stone, Alberta, Canada, *Journal of Field Archaeology* 6:41–48.

Keyser, J.D., and Davis, C.M., 1982, Ceramics from the Highwalker Site: A Study of Late Prehistoric Period Cultural Systematics on the Northwestern Plains, *Plains Anthropologist* 27:287–303.

Kiltie, R., 1984, Seasonality, Gestation Time, and Large Mammal Extinctions, in: *Quaternary Extinctions: A Prehistoric Revolution* (P. Martin and R. Klein, eds.), Univeristy of Arizona Press, Tucson, pp. 299–314.

Kincer, J.B., 1923, The Climate of the Great Plains as a Factor in Their Utilization, *Annals of the Association of American Geographers* 13:67–80.

Knudson, R., 1983, Organizational Variability in Late Paleoindian Assemblages, *Washington State University Laboratory of Anthropology Report of Investigations* 60.

Koford, C.B., 1957, The Vicuna and the Puna, *Ecological Monographs* 27:152–219.

Krieger, A.D., 1964, Early Man in the New World, in: *Prehistoric Man in the New World* (J.D. Jennings and E. Norbeck, eds.), University of Chicago Press, Chicago, pp. 23–84.

Kroeber, A.L., 1908, Ethnology of the Gros Ventre, *American Museum of Natural History Anthropological Papers* Vol. 1, Pt. 4.

Kroeber, A.L., 1939, Cultural and Natural Areas of Native North America, *University of California Publications in American Archaeology and Ethnology* 38.

Kroeber, A.L., 1983, *The Arapahoe,* Garland Publishing, New York.

Kutzbach, J.E., 1976, The Nature of Climate and Climatic Variations, *Quaternary Research* 6:471–480.

Kvamme, K., 1984, *New Methods for Investigating the Environmental Basis of Prehistoric Site Locations,* PhD dissertation, Anthropology Department, University of California, Santa Barbara.

Lamb, H.H., 1966, *The Changing Climate,* Methuen and Co., London.

Lamb, H.H., 1977, *Climate Present, Past, and Future Vol. 2: Climatic History and the Future,* Methuen and Co., London.

Lamb, H.H., 1981, An Approach to the Study of the Development of Climate and Its Impact in Human Affairs, in: *Climate and History* (T.M.L. Wigley, M.J. Ingram, and G. Farmer, eds.), Cambridge University Press, New York, pp. 291–309.

Latady, W.R., 1985, *Bison Foraging Possibilities in the Washakie Basin of Wyoming,* Paper Presented at the Forty-Third Annual Plains Conference, Iowa City, Iowa.

Laws, R.M., Parker, I.S.C., and Johnstone, R.C.B., 1975, *Elephants and Their Habitats,* Clarendon Press, Oxford.

Lawson, P.L., 1974, *The Climate of the Great American Desert,* University of Nebraska Studies, New Series No. 46.

Lee, R.B., 1968, What Hunters Do for a Living, or How to Make Out on Scarce Resources, in: *Man the Hunter* (R.B. Lee and I. DeVore, eds.), Aldine, Chicago, pp. 30–48.

Levins, S.A., and Paine, R.T., 1974, Disturbance, Patch Formation, and Community Structure, *Proceedings of the National Academy of Sciences* 71:2744–2747.

Levy, J.E., 1961, Ecology of the South Plains, in: Patterns of Land Utilization and Other Papers (V. Garfield, ed.), *Proceedings of the American Ethnological Society* 18:18–26.

Lewis, O., 1942, The Effects of White Contact upon Blackfoot Indian Culture, *Monographs of the American Ethnological Society* 6.

Longley, R.W., 1974, Spatial Variation of Precipitation over the Canadian Prairies, *Monthly Weather Review* 102:307–312.

Lorrain, D., 1968, Analysis of the Bison Bones, in: Bonfire Shelter: A Stratified Bison Kill Site in the Amistad Reservoir Area, Val Verde County, Texas, by D. Dibble, *Texas Archaeological Salvage Project Miscellaneous Papers* 5, pp. 91–119.

Lott, D.F., 1979, Dominance Relations and Breeding Rate in Mature Male American Bison, *Zeitgeist für Tierpsychologie* 49:418–432.

Lott, D.F., and Minta, S.C., 1983, Random Individual Association and Social Group Instability in American Bison (Bison bison), *Zietschrift für Tierpsychologie* 61:153–172.

Lowie, R.H., 1909a, The Assiniboine, *American Museum of Natural History Anthropological Papers* Vol. 4, pt. 1.

Lowie, R.H., 1909b, The Northern Shoshone, *American Museum of Natural History Anthropological Papers* 2:165–306.

Lowie, R.H., 1916a, Military Societies of the Crow Indians, *American Museum of Natural History Anthropological Papers* 11:143–218.

Lowie, R.H., 1916b, Societies of the Kiowa, *American Museum of Natural History Anthropological Papers* 11:837–852.

Lowie, R.H., 1916c, Plains Indian Age-Societies: Historical and Comparative Summary, *American Museum of Natural History Anthropological Papers* 11:877–986.

Lowie, R.H., 1935, *The Crow Indians,* Farrar and Rinehart, New York.

Lowie, R.H., 1954, *Indians of the Plains,* Natural History Press, Garden City, New York.

Lyttleton, J.W., 1973, Proteins and Nucleic Acids, in: *Chemistry and Biochemistry of Herbage* (G.W. Butler and R.W. Bailey, eds.), Academic Press, New York, pp. 63–105.

Macy, P., 1936, The Quantitative Mineral Nutrient Requirements of Plants, *Plant Physiology* 11:749–764.

Maddock, L., 1979, The "Migration" and the Grazing Succession, in: *Serengeti: Dynamics of an Ecosystem,* (A.R.E. Sinclair and M. Norton-Griffiths, eds.), University of Chicago Press, Chicago, pp. 103–129.

Mandelbaum, D.G., 1940, The Plains Cree, *American Museum of Natural History Anthropological Papers* No. 37, pt. 2.

Marmaduke, W.S., and Whitsett, H., 1975, Reconnaissance and Archaeological Studies in the Canadian River Valley, in: *Canadian Breaks* (D. Kennard, ed.), pp. 68–126. Division of Natural Resources and Environment, University of Texas, Austin.

Martin, A.C., Zim, H.A., and Nelson, A.L., 1961, *American Wildlife and Plants,* Dover, New York.

Martin, P., 1973, The Discovery of America, *Science* 179:969–974.

Martin, P., 1984, Prehistoric Overkill: The Global Model, in: *Quaternary Extinctions: A Prehistoric Revolution* (P. Martin and R. Klein, eds.), University of Arizona Press, Tucson, pp. 354–403.

Mather, J.R., 1964, Average Climatic Water Balance of the Continents, Part 6: North America (Excluding United States) and Part 7: United States. *C.W. Thornthwaite Associates Laboratory of Climatology Publications in Climatology* Vol. 17, Nos. 2 and 3.

Mayall, M.P., 1971, *The Kiowas,* 2nd Edition, University of Oklahoma Press, Norman.

McCartney, P.H., 1985, *Alternative Hunting Strategies in Plains Paleoindian Adaptations,* Paper read at the Annual Meeting of the Society for American Archaeology, Denver, Colorado.

McDaniel, K.C., 1978, *Evaluation of LANDSAT MSS Data for Classifying and Characterizing Natural Vegetation on a Regional Basis,* PhD Dissertation, Texas A and M University.

McDonald, J.N., 1981, *North American Bison,* University of California Press, Berkeley.

McDonald, J.N., 1984, The Reordered North American Selection Regime and Late Quaternary Megafaunal Extinctions, in: *Quaternary Extinctions* (P. Martin and R. Klein, eds.), University of Arizona Press, Tucson, pp. 404–439.

McGuire, R.H., 1983, Breaking Down Cultural Complexity: Inequality and Heterogeneity, in: *Advances in Archaeological Method and Theory* (M. Schiffer, ed.), 6:91–142.

McHugh, T., 1958, Social Behavior of the American Buffalo (Bison Bison Bison), *Zoologica* 43:1–40.

McHugh, T., 1972, *The Time of the Buffalo,* Alfred A. Knopf, New York.

McLaughlin, D., n.d., Preliminary Thoughts on the Canyon Lakes 6 Core, manuscript in author's possession.

McMillan, R.B., and Klippel, W.E., 1981 Post-Glacial Environmental Change and Hunting-Gathering Societies of the Southern Prairie Peninsula, *Journal of Archaeological Science* 8:215–245.

McNaughton, S.J., 1976, Serengeti Migratory Wildebeest: Facilitation of Energy Flow by Grazing, *Science* 191:92–94.

McNaughton, S.J., 1979, Grassland Herbivore Dynamics, in: *Serengeti: Dynamics of an Ecosystem* (A.R.E. Sinclair and M. Norton-Griffiths, eds.), University of Chicago Press, Chicago, pp. 46–81.

McNaughton, S.J., 1984, Grazing Lawns: Animals in Herds, Plant Form, and Coevolution, *American Naturalist* 124:863–886.
Meagher, M.M., 1973, The Bison of Yellowstone Park, *National Park Service Monograph Series* 1.
Meissner, H.H., 1982, Theory and Application of a Method to Calculate Forage Intake of Wild Southern African Ungulates for Purposes of Estimating Carrying Capacity, *South African Journal of Wildlife Research* 112:41–47.
Melton, F.A., 1940, A Tentative Classification of Sand Dunes—Its Application to Dune History in the Southern High Plains, *Journal of Geology* 48:113–174.
Meltzer, D., and Mead, J., 1983, The Timing of Late Pleistocene Mammalian Extinctions in North America, *Quaternary Research* 19:130–135.
Mentis, M.T., 1977, Stocking Rates and Carrying Capacities for Ungulates on African Grasslands, *South African Journal of Wildlife Research* 7:89–98.
Mentis, M.T., and Duke, F., 1976, Carrying Capacities of Natural Veld in Natal for Large Wild Herbivores, *South African Journal of Wildlife Research* 6:65–74.
Mloszewski, M.J., 1983, *The Behavior and Ecology of the African Buffalo,* Cambridge University Press, Cambridge.
Monson, R.K., and Williams, III, G.J., 1982, A Correlation between Photosynthetic Temperature Adaptation and Seasonal Phenology Patterns in the Shortgrass Prairie, *Oecologia* 54:58–62.
Moodie, D.W., and Ray, A.J., 1976, Buffalo Migrations in the Canadian Plains, *Plains Anthropologist* 21:45–52.
Mooney, J., 1895, Calendar History of the Kiowa Indians, *Annual Report of the Bureau of American Ethnology* 17(1).
Mooney, J., 1907, The Cheyenne Indians, *American Anthropological Association Memoirs* Vol. 1, Part 6.
Moore, J.H., 1974, Cheyenne Political History, 1820–1894, *Ethnohistory* 21:329–360.
Moore, J.H., 1981, Evolution and Historical Reductionism, *Plains Anthropologist* 26:261–270.
Morgan, L.H., 1964, *Ancient Society,* Harvard University Press, Cambridge.
Morgan, R.G., 1980, Bison Movement Patterns on the Canadian Plains: An Ecological Analysis, *Plains Anthropologist* 25:143–160.
Moser, L.E., 1977, Carbohydrate Translocation in Range Plants, in: *Rangeland Plant Physiology* (R.E. Sosebee, ed.), Society for Range Management, Denver, pp. 47–72.
Mosiman, J.E., and Martin, P., 1975, Simulating Overkill by Paleoindians, *American Scientist* 63:304–313.
Moss, R., Watson, A., and Ollason, J., 1982, *Animal Population Dynamics,* Chapman and Hall, New York.
Mulloy, W., 1942, The Hagen Site, *University of Montana Publications in the Social Sciences* 1.
Mulloy, W., 1958, A Preliminary Historical Outline for the Northwestern Plains, *University of Wyoming Publications, Publications in Science* 22.
Murray, M.G., 1982, Home Range, Dispersal, and the Clan System of Impala, *African Journal of Ecology* 20:253–269.
Murton, R.K., 1971, Why Do Some Bird Species Feed in Flocks? *Ibis* 113:534–536.
Naroll, R., 1956, A Preliminary Index of Social Development, *American Anthropologist* 58:687–715.
National Research Council, 1971, *Atlas of Nutritional Data on United States and Canadian Feeds,* National Academy of Sciences, Washington, DC.
Newbauer, J.J., White, L.H., Moy, R.M., and Perry, D.A., 1980, Effects of Increased Rainfall on Native Range Production in Eastern Montana, *Journal of Range Management* 33:246–250.
Nials, F., 1980, Environment, in: Implications of Archaeological Collections, Test, and Excavations in the Carlsbad Area (,M. Thompson, ed.), *New Mexico State University Cultural Resources Management Division Report* 433, pp. 4–23.
Ode, D.J., Tieszen, L.L., and Lerman, J.C., 1980, The Seasonal Contribution of C3 and C4 Plant Species to Primary Production in a Mixed Prairie, *Ecology* 61:1304–1311.
Odum, H.T., 1971, *Environment, Power, and Society,* John Wiley and Sons, New York.

Oldfield, F., 1975, Pollen-Analytical Results, Part II, in: Late Pleistocene Environments of the Southern High Plains (F. Wendorf and J. Hester, eds.), *Fort Burgwin Research Center Publication* 9, pp. 121–148.
Oldfield, F., and Schoenwetter, J., 1975, Discussion of the Pollen-Analytical Evidence, in: Late Pleistocene Environments of the Southern High Plains (F. Wendorf and J. Hester, eds.), *Fort Burgwin Research Center Publication* 9, pp. 149–178.
Oliver, S.C., 1962, Ecology and Cultural Continuity as Contributing Factors in the Social Organization of the Plains Indian, *University of California Publications in American Archaeology and Ethnology* 48:1–90.
Osborn, A.J., 1983, Ecological Aspects of Equestrian Adaptations in North America, *American Anthropologist* 85:563–591.
Peden, D.G., 1972, *The Trophic Relations of Bison Bison to the Shortgrass Plains,* PhD Dissertation, Colorado State University, Fort Collins.
Pennycuick, C.J., 1979, Energy Costs of Locomotion and the Concept of "Foraging Radius," in: *Serengeti: Dynamics of an Ecosystem* (A.R.E. Sinclair and M. Norton-Griffiths. eds.), University of Chicago Press, Chicago, pp. 164–184.
Pennycuick, L., 1975, Movements of Migratory Wildebeest Population in the Serengeti Area between 1960 and 1973, *East African Wildlife Journal* 13:65–87.
Perry, R.J., 1980, The Apachean Transition from the Subactic to the Southwest, *Plains Anthropologist* 25:279–296.
Pettit, R.D., and Deering, D.W., 1974, Yield and Protein Content of Sandyland Range Forages as Affected by Three Nitrogen Fertilizers, *Journal of Range Management* 27:211–213.
Pianka, E.R., 1970, On *r*- and *K*-Selection, *American Naturalist* 104:592–597.
Pianka, E.R., 1974, *Evolutionary Ecology,* Harper and Row, New York.
Pyke, G.H., 1984, Optimal Foraging theory: A Critical Review, *Annual Review of Ecology and Systematics* 15:523–575.
Pyke, G.H., Pulliam, H.R., and Charnov, E.L., 1977, Optimal Foraging: A Critical Review of Theory and Tests, *Quarterly Review of Biology* 52:137–154.
Rancier, J., Haynes, G., and Stanford, D., 1982, 1981 Investigations at the Lamb Spring Site, *Southwestern Lore* 48:1–17.
Rannie, W.F., 1983, Breakup and Freezeup of the Red River at Winnepeg, Manitoba Canada in the 19th Century and Some Climatic Implications, *Climatic Change* 5:283–296.
Rauzi, F., and Dobrenz, A.K., 1970, Seasonal Variation of Chlorophyll in Western Wheatgrass and Blue Grama, *Journal of Range Management* 23:372–373.
Rauzi, F., Painter, L.J., and Dobrenz, A.K., 1969, Mineral and Protein Contents of Blue Grama and Western Wheatgrass, *Journal of Range Management* 22:47–49.
Ray, A.J., 1974, *Indians in the Fur Trade,* University of Toronto Press, Toronto.
Read, D., 1983, *Global Features of the Spatial Distribution of Great Basin Sites in the MX Project Area,* Paper read at the annual meeting of the Society for American Archaeology, Pittsburgh, Pennsylvania.
Reeves, C.C., 1965a, Chronology of West Texas Pluvial Lake Dunes, *Journal of Geology* 73:504–508.
Reeves, C.C., 1965b, Pleistocene Climate of the Llano Estacado, *Journal of Geology* 73:181–189.
Reeves, C.C., 1966a, Pluvial Lake Basins of West Texas, *Journal of Geology* 74:269–291.
Reeves, C.C., 1966b, Pleistocene Climate of the Llano Estacado, II, *Journal of Geology* 74:642–647.
Reeves, C.C., 1976, Quaternary Stratigraphy and Geologic History of Southern High Plains, Texas and New Mexico, in: *Quaternary Stratigraphy of North America* (W.C. Muhaney, ed.), Dowden, Hutchinson, and Ross, Inc., Stroudsberg, PA, pp. 213–235.
Reeves, C.C., and Parry, W.T., 1965, Geology of West Texas Pluvial Lake Carbonates, *American Journal of Science* 263:606–615.
Reeves, C.C., and Parry, W.T., 1969, Age and Morphology of Small Lake Basins, Southern High Plains, Texas and Eastern New Mexico, *Texas Journal of Science* 20:349–354.

Reher, C.A., 1974, Population Study of the Casper Site Bison, in: *The Casper Site* (G. Frison, ed.), Academic Press, New York, pp. 113–124.
Reher, C.A., 1978, *Adaptive Process on the Late Prehistoric Short Grass Plains,* Ph.D. Dissertation, Anthropology Department, University of New Mexico.
Reher, C., and Frison, G.C., 1980, The Vore Site, 48CK302, a Stratified Buffalo Jump in the Wyoming Black Hills, *Plains Anthropologist Memoir* 16.
Reuss, J.O., and Innis, G.S., 1978, A Grassland Nitrogen-Flow Simulation Model, in: *Grassland Simulation Model* (G.S. Innis, ed.), Springer-Verlag, New York, pp. 185–204.
Reynolds, H.W., Hansen, R.M., and Peden, D.G., 1978, Diets of the Slave River Lowland Bison Herd, Northwest Territories, Canada, *Journal of Wildlife Management* 42:581–590.
Roberts, F.H.H., 1942. Archaeological and Geological Investigations in the San Jon District, Eastern New Mexico, *Smithsonian Miscellaneous Collections* 103(4).
Roe, F.G., 1970, *The North American Buffalo,* 2nd edition, University of Toronto Press, Toronto.
Rovner, I., and Agogino, G., 1967, An Analysis of Fluted and Unfluted Folsom Points from Blackwater Draw, *The Masterkey* 41:131–137.
Rutberg, A.T., 1983, Factors lnfluencing Dominance Status in American Bison Cows (Bison bison), *Zietschrift für Tierpsychologie* 63:206–216.
Sala, O.E., and Lauenroth, W.K., 1982, Small Rainfall Events: an Ecological Role in Semiarid Regions, *Oecologia* 53:301–304.
Sampson, A.W., 1952, *Range Management: Principles and Practices,* John Wiley and Sons, New York.
Sanchez, W.A., and Kutzbach, J.E., 1974, Climate of the American Tropics and Subtropics in the 1960s and Possible Comparisons with Climatic Variations of the Last Millennium, *Quaternary Research* 4:128–135.
Saunders, J.J., 1980, A Model for Man-Mammoth Relationships in Late Pleistocene North America, *Canadian Journal of Anthropology* 1:87–98.
Schalk, R.F., 1979, Land Use and Organizational Complexity among Foragers of Northwestern North America, in: Affluent Foragers: Pacific Coasts East and West (S. Koyama and D.H. Thomas, eds.), *Senri Ethnological Studies* 9, pp. 53–76.
Schoenwetter, J., 1975, Pollen-Analytical Results, Part I, in: *Late Pleistocene Environments of the Southern High Plains* (F. Wendorf and J. Hester, eds.), Fort Burgwin Research Center Publication No. 9, pp. 103–120.
Schoolcraft, H.W. 1851–1858, *Information Respecting the History, Condition, and Prospects of the Indian Tribes of the United States Vols. 1–5,* J. Lippincott, Philadelphia.
Schroeder, A.H., 1974, *Apache Indians I,* Garland Publishing Co, New York.
Schultz, C.B., 1943, Some Artifact Sites of Early Man in the Great Plains, *American Antiquity* 8:242–249.
Secoy, F., 1951, The Identity of the "Paduca," *American Anthropolgist* 53:525–542.
Secoy, F., 1953, Changing Military Patterns on the Great Plains (17th Century through Early 19th Century), *American Ethnological Society Monograph* 21.
Sellards, E.H., 1952, *Early Man in America,* University of Texas Press, Austin.
Sellards, E.H., 1955, Fossil Bison and Associated Artifacts from Milnesand, New Mexico, *American Antiquity* 20:336–344.
Sellards, E., Evans, G., and Meade, G., 1947, Fossil Bison and Associated Artifacts from Plainview, Texas, *Bulletin of the Geological Society of America* 58:927–964.
Service, E.R., 1962. *Primitive Social Organization,* Random House, New York.
Service, E.R., 1966, *The Hunters,* Prentice-Hall, Englewood Cliffs, NJ.
Shackleton, D.M., 1968, *Comparative Aspects of Social Organization of American Bison,* Master's Thesis, University of Western Ontario.
Sharrock, S.R., 1974, Cree, Cree-Assiniboines, and Assiniboines: Interethnic Social Organization on the Far Northern Plains, *Ethnohistory* 21:95–122.

Shay, C.T., 1971, *The Itasca Bison Kill Site: An Ecological Analysis,* Minnesota Historical Society.

Shimkin, D.B., 1941, Shoshone-Comanche Origins and Migrations, *Proceedings of the Sixth Pacific Science Conference* 4:17–25.

Sims, P.L., and Singh, J.S., 1978a, The Structure and Function of Ten Western North American Grasslands II: Intra-Seasonal Dynamics in Primary Producer Compartments, *Journal of Ecology* 66:547–572.

Sims, P.L., and Singh, J.S., 1978b, The Structure and Function of Ten Western North American Grasslands III: Net Primary Production, Turnover and Efficiencies of Energy Capture and Water Use, *Journal of Ecology* 66:573–597.

Sims, P.L., Singh, J.S., and Lauenroth, W.K., 1978, The Structure and Function of Ten North American Grasslands I: Abiotic and Vegetational Characteristics, *Journal of Ecology* 66:251–285.

Sinclair, A.R.E., 1975, The Resource Limitation of Trophic Levels in Tropical Grassland Ecosystems, *Journal of Animal Ecology* 44:497–520.

Sinclair, A.R.E., 1977, *The African Buffalo,* University of Chicago Press, Chicago.

Sinclair, A.R.E., 1979, Dynamics of the Serengeti Ecosystem: Process and Pattern, in: *Serengeti: Dynamics of an Ecosystem* (A.R.E. Sinclair and M. Norton-Griffiths, eds.), University of Chicago Press, Chicago, pp. 1–30.

Sinclair, A.R.E., and Norton-Griffiths, M., 1982, Does Competition or Facilitation Regulate Migrant Ungulate Populations in the Serengeti? A Test of Hypotheses, *Oecologia* 53:364–369.

Skinner, A., 1916, Political Organization, Cults, and Ceremonies of the Plains Cree, *American Museum of Natural History Anthropological Papers* 11, pp. 513–542.

Smiley, F.E., 1978, Changes in the Cursorial Ability of Wyoming Holocene Bison, *Wyoming Contributions to Anthropology* 1:105–126.

Smith, C., Runyon, J., and Agogino, G., 1966, A Progress Report on a Pre-Ceramic Site at Rattlesnake Draw, Eastern New Mexico, *Plains Anthropologist* 11:302–313.

Smith, E.A., 1979, Human Adaptation and Energetic Efficiency, *Human Ecology* 7:53–74.

Smith, E.A., 1981, The Application of Optimal Foraging Theory to the Analysis of Hunter-Gatherer Group Size, in: *Hunter-Gatherer Foraging Strategies* (B. Winterhalder and E.A. Smith, eds.), University of Chicago Press, Chicago, pp. 36–65.

Smith, G.I., and Smith-Perrott, F.A., 1983, Pluvial Lakes of the Western United States, in: *Late Quaternary Environments of the United States Vol 1: The Late Pleistocene* (S.C. Porter, ed.), University of Minnesota Press, Minneapolis.

Sneva, F.A., 1976, Correlations of Precipitation and Temperature with Spring Regrowth, and Mature Crested Wheatgrass Yields, *Journal of Range Management* 30:270–275.

Sollberger, J.L., 1986, Discussion of the Akerman and Fagan Paper, *Lithic Technology* 15:7.

Soper, J.D., 1941, History, Range, and Home Life of the Northern Bison, *Ecological Monographs* 11:348–412.

Speer, R., 1980, *Overview Assessment of the Archaeological Resources of Eleven Counties in the Northwestern Texas Panhandle,* manuscript in author's possession.

Speth, J.D., 1983, *Bison Kills and Bone Counts,* University of Chicago Press, Chicago.

Speth, J.D., and Spielmann, K.A., 1983, Energy Source, Protein Metabolism, and Hunter-Gatherer Subsistence Strategies, *Journal of Anthropological Archaeology* 2:1–31.

Spielmann, K.A., 1983, Late Prehistoric Exchange between the Southwest and Southern Plains, *Plains Anthropologist* 28:257–272.

Stafford, T.W., 1981, Alluvial Geology and Archaeological Potential of the Southern High Plains, *American Antiquity* 46:548–565.

Stahle, D., and Cleaveland, M., 1988, Texas Drought History Reconstructed and Analyzed from A.D. 1698 to 1980, *Journal of Climate,* in press.

Stanford, D., 1978, The Jones-Miller Site: An Example of Hell Gap Bison Procurement Strategy, in: Bison Procurement and Utilization: A Symposium (L. Davis and M. Wilson, eds.), *Plains Anthropologist Memoir* 16, pp. 90–97.

Steward, J.H., 1938, Basin-Plateau Aboriginal Sociopolitical Groups, *Bureau of American Ethnology Bulletin* 120.

Stewart, F.H., 1977, *Fundamentals of Age-Group Systems,* Academic Press, New York.

Stockton, C.W., and Meko, D.M., 1983, Drought Recurrence in the Great Plains as Reconstructed from Long-Term Tree-Ring Records, *Journal of Climate and Applied Meteorology* 22:17–29.

Suttles, W., 1968, Coping With Abundance: Subsistence on the Northwest Coast, in: *Man the Hunter* (R.B. Lee and I. DeVore, eds.). Aldine, Chicago, pp. 56–68.

Sutton, M., 1986, Warfare and Expansion: An Ethnohistoric Perspective on the Numic Spread, *Journal of California and Great Basin Anthropology* 8:65–82.

Syms, E.L., 1977, Cultural Ecology and Ecological Dynamics of the Ceramic Period in Southwestern Manitoba, *Plains Anthropologist Memoir* 12.

Talbot, L.M., 1962, Food Preferences of Some East African Wild Ungulates, *East African Agricultural and Forestry Journal* 27:131–138.

Talbot, L., and Talbot. M., 1963a, The High Biomass of Wild Ungulates on East African Savannah, *Transactions of the North American Wildlife and Natural Resources Conference* 28:465–476.

Talboy, L., and Talbot, M., 1963b, The Wildebeest in Western Masailand, *Wildlife Monographs* 12.

Teer, J.G., Thomas, J.W., and Walker, E.A., 1965, Ecology and Management of White-Tailed Deer in the Llano Basin of Texas, *Wildlife Monographs* 12.

Teeri, J.A., and Stowe, L.G., 1976, Climatic Patterns and the Distribution of C4 Grasses in North America, *Oecologia* 23:1–12.

Teit, J.A., 1930, The Salishan Tribes of the Western Plateaus, *Bureau of American Ethnology Annual Report* 45:25–396.

Thomas, A.B., 1935, *After Coronado: Spanish Exploration Northeast of New Mexico, 1696–1727,* University of Oklahoma Press, Norman.

Thomas, D.H., 1972, Western Shoshone Ecology, in: Great Basin Cultural Ecology: A Symposium (D.D. Fowler, ed.), *Desert Research Institute Publications in the Social Sciences* 8, pp. 135–153.

Thomas, R.B., Winterhalder, B., and McCrae, S.D., 1979, An Anthropological Approach to Human Ecology and Adaptive Dynamics, *Yearbook of Physical Anthropology* 22:1–46.

Thompson, L.M., and Troeh, F.R., 1973, *Soils and Soil Fertility,* McGraw-Hill Book Company, New York.

Thornthwaite, C.W., 1941, Climate and Settlement in the Great Plains, *Yearbook of Agriculture* 1941:177–187.

Todd, L.C., 1983, *The Horner Site: Taphonomy of an Early Holocene Bison Bonebed,* PhD Dissertation, Anthropology Department, University of New Mexico.

Tomanek, G.W., 1959, The Effects of Climate and Grazing on Mixed Prairie, in: *Grasslands* (H.B. Sprague, ed.), American Association for the Advancement of Science, Washington, pp. 371–378.

Torrence, R., 1983, Time Budgeting and Hunter-Gatherer Technology, in: *Pleistocene Hunter-Gatherers in Europe* (G. Bailey, ed.), University of Cambridge Press, Cambridge, pp. 11–22.

Trenholm, V.C., 1970, *The Arapahoes, Our People,* University of Oklahoma Press, Norman.

Trlica, M.J., 1977, Distribution and Utilization of Carbohydrate Reserves in Range Plants, in: *Rangeland Plant Physiology* (R.E. Sosebee, ed.), Society for Range Management, Denver, pp. 73–96.

Tunnell, C., 1977, Fluted Point Production as Revealed by Lithic Specimens from the Adair-Steadman Site in Northwest Texas, in: *Paleoindian Lifeways* (E. Johnson, ed.), West Texas Museum Association, Lubbock, TX, pp. 140–168.

United States Commissioner of Indian Affairs, 1869, *Annual Report,* Bureau of the Interior, Washington DC.

Van Soest, P.J., 1982, *Nutritional Ecology of the Ruminant,* O and B Books, Corvallis, Oregon.

Vines, R.G., 1982, Rainfall Patterns in the Western United States, *Journal of Geophysical Research* 87:7303–7311.

Wahl, E.W., and Lawson. T.L., 1970, The Climate of the Midnineteenth Century United States Compared to the Current Normals, *Monthly Weather Review* 98:259–265.

Walker, D.N., and Boyce, M.S., 1985, *Multivariate Allometry and the Evolution of the North American Bison,* Paper read at the Plains Antrhopological Conference, Iowa City, Iowa.

Walker, J., 1982, *Lakota Society* (edited by R. DeMaillie), University of Nebraska Press, Lincoln.

Wallace, E., and Hoebel, E.A., 1952, *The Comanches,* University of Oklahoma Press, Norman.

Warnica, J.M., 1961, The Elida Site, Evidence of a Folsom Occupation in Roosevelt County, Eastern New Mexico, *Bulletin of the Texas Archaeological Society* 30:209–215.

Warnica, J., and Williamson, T., 1968, The Milnesand Site—Revisited, *Transactions of the Third Regional Archaeological Symposium for Southeastern New Mexico and Western Texas* pp. 21–22.

Warren, G.K., 1856, Explorations in the Dacota Country, in the Year 1855, *Senate Executive Document* 76.

Weaver, J.E., 1954, *North American Prairie,* Johnsen Publishing Company, Lincoln, Nebraska.

Weaver, J.E., and Albertson, F.W., 1956, *Grasslands of the Great Plains,* Johnsen Publishing Company, Lincoln, Nebraska.

Weaver, J.E., and Clements, F.E., 1938, *Plant Ecology,* McGraw-Hill, New York.

Weaver, J.E., and Tomanek, G.W., 1951, Ecological Studies in a Midwestern Range: The Vegetation and Effects of Cattle on Its Composition and Distribution, *Nebraska Conservation Service Bulletin* 31.

Wedel, W.R., 1941, Environment and Native Subsistence Economies in the Central Great Plains, *Smithsonian Miscellaneous Collections* 101(3).

Weinmann, H., 1955, The Chemistry and Physiology of Grasses, in: *The Grasses and Pastures of South Africa* (D. Meredith, ed.), Central News Agency, Johannesburg, pp. 571–600.

Weist, K.M., 1980, Plains Indian Women: An Assessment, in: *Anthropology on the Great Plains* (W.R. Wood and M. Liberty, eds.), University of Nebraska Press, Lincoln, pp. 255–271.

Wendorf, F., 1970, The Lubbock Subpluvial, in: Pleistocene and Recent Environments of the Great Plains (W. Dort and J.K. Jones, eds.), *University of Kansas Department of Geology Special Publication* 3, pp. 23–36.

Wendorf, F., 1975a, The Modern Environment, in: *Late Pleistocene Environments of the Southern High Plains,* (F. Wendorf and J. Hester, eds.), Fort Burgwin Research Center Publication 9, pp. 1–12.

Wendorf, F., 1975b, Summary and Conclusions, in: *Late Pleistocene Environments of the Southern High Plains,* (F. Wendorf and J. Hester, eds.), Fort Burgwin Research Center Publication 9, pp. 257–279.

Wendorf, F., and Hester, J., 1962, Early Man's Use of the Great Plains Environment, *American Antiquity* 28:159–171.

Wendorf, F., and Krieger, A., 1959, New Light on the Midland Discovery, *American Antiquity* 25:66–78.

Wendorf, F., Krieger, A.D., and Albritton, C., 1955, *The Midland Discovery,* University of Texas Press, Austin.

Western, D., 1975, Water Availability and Its Influence on the Structure and Dynamics of a Savannal Large Mammal Community, *East African Wildlife Journal* 13:265–286.

Western, D., 1979, Size, Life History, and Ecology in Mammals, *African Journal of Ecology* 17:185–204.

Wheat, J.B., 1972, The Olsen-Chubbock Site: A Paleo-Indian Bison Kill, *Society for American Archaeology Memoir* 26.

Wheat, J.B., 1979, The Jurgens Site, *Plains Anthropologist Memoir* 15.

White, L.M., 1973, Carbohydrate Reserves of Grasses: A Review, *Journal of Range Management* 26:13–18.

Wiens, J., 1976, Population Responses to a Patchy Environment, *Annual Review of Ecology and Systematics* 7:81–120.

Willard, E.E., and Schuster, J.L., 1973, Chemical Composition of Six Southern Great Plains Grasses as Related to Season and Precipitation, *Journal of Range Management* 26:37–38.

Wilmsen, E., 1973, Interaction, Spacing Behavior, and the Organization of Hunting Bands, *Journal of Anthropological Research* 29:1–31.

Wilmsen, E., and Roberts, F.H.H., 1978, Lindenmeier, 1934–1974: Concluding Report on Investigations, *Smithsonian Contributions in Anthropology* 24.

Wilson, M., 1969, Problems in the Speciation of American Fossil Bison, in: *Post-Pleistocene Man and His Environment on the Northern Plains* (M. Wilson, ed.), University of Calgary Archaeological Association, pp. 177–197.

Wilson, M., 1974, Fossil Bison and Artifacts from the Mona Lisa Site, Calgary, Alberta, Part 1: Stratigraphy and Artifacts, *Plains Anthropologist* 19:34–45.

Wilson, M., 1978, Archaeological Kill Site Populations and the Holocene Evolution of the Genus Bison, in: *Bison Procurement and Utilization: A Symposium* (L. Davis and M. Wilson, eds.), *Plains Anthropologist Memoir* 14, pp. 9–22.

Winterhalder, B., 1980, Environmental Analysis in Human Evolution and Adaptation Research, *Human Ecology* 8:135–170.

Winterhalder, B., 1981, Optimal Foraging Strategies and Hunter-Gatherer Research in Anthropology: Theory and Models, in: *Hunter-Gatherer Foraging Strategies* (B. Winterhalder and E.A. Smith, eds.), University of Chicago Press, Chicago, pp. 11–35.

Winterhalder, B., and Smith, E.A., 1981, *Hunter-Gatherer Foraging Strategies,* University of Chicago Press, Chicago.

Wissler, C., 1910, Material Culture of the Blackfoot Indians, *American Museum of Natural History Anthropological Papers* 5, pp. 1–176.

Wissler, C., 1912a, Social Organization and Ritualistic Ceremonies of the Blackfoot Indians, *American Museum of Natural History Anthropological Papers* 7.

Wissler, C., 1912b, The Social Life of the Blackfoot Indians, *American Museum of Natural History Anthropological Papers* 7, pp. 1–64.

Wissler, C., 1916a, Societies and Ceremonial Associations of the Oglala Division of the Teton Dakota, *American Museum of Natural History Anthropological Papers* 11, pp. 1–100.

Wissler, C., 1916b, Societies and Dance Associations of the Blackfoot Indians, *American Museum of Natural History Anthropological Papers* 11, pp. 359–460.

Wobst, H.M., 1974, Boundary Conditions for Paleolithic Social Systems: A Simulation Approach, *American Antiquity* 39:147–177.

Wood, W.R., and Downer, A.S., 1977, Notes on the Crow-Hidatsa Schism, *Plains Anthropologist* 22:83–100.

Wormington, H.M., 1957, Ancient Man in North America, *Denver Museum of Natural History Popular Series* 4.

Wright, G.A., 1978, The Shoshonean Migration Problem, *Plains Anthropologist* 23:113–137.

Wright, H.E., 1970, Vegetational History of the Central Plains, in: Pleistocene and Recent Environments of the Great Plains (W. Dort and J.K. Jones, eds.), *University of Kansas Department of Geology Special Publication* 3, pp. 157–172.

Yellen, J., and Lee, R.B., 1976. The Dobe-/Du/da Environment, in: *Kalahari Hunter-Gatherers* (R.B. Lee and I. Devore, eds.), Harvard University Press, Cambridge, pp. 27–46.

Index